Notetaking Guide Teacher's Edition

The Notetaking Guide contains a lesson-by-lesson framework that allows students to take notes on and review the main concepts of each lesson in the textbook. Each Notetaking Guide lesson features worked-out examples and Checkpoint exercises. Each example has a number of write-on lines for students to complete, either in class as the example is discussed or at home as part of a review of the lesson. Each chapter concludes with a review of the main vocabulary of the chapter. Upon completion, each chapter of the Notetaking Guide can be used by students to help review for the test on that particular chapter.

The Teacher's Edition contains annotated overprinted answers for all write-on lines and exercises. In addition, the Notetaking Guide can be used in conjunction with the Notetaking Guide Transparencies, which are available separately.

McDougal Littell
A HOUGHTON MIFFLIN COMPANY
Evanston, Illinois • Boston • Dallas

ISBN: 0-618-41061-9

3456789–MDO–07 06 05 04

Contents

Algebra 1 Concepts and Skills Notetaking Guide

Contents

Contents

Contents

1.1 Variables in Algebra

Goal Evaluate variable expressions.

VOCABULARY

Variable A variable is a letter used to represent a range of numbers.

Values The numbers a variable represents are called values of the variable.

Variable expression A variable expression consists of constants, variables, and operations.

Numerical expression An expression that represents a particular number is called a numerical expression.

Evaluate To evaluate a variable expression, you write the expression, substitute a number for each variable, and simplify.

Example 1 ***Describe the Variable Expression***

The multiplication symbol × is usually not used in algebra because of its possible confusion with the variable *x*.

Variable Expression	Meaning	Operation
a. $3x$, $3 \cdot x$, $(3)(x)$	3 times x	Multiplication
b. $\frac{14}{y}$, $14 \div y$	14 divided by y	Division
c. $9 + a$	9 plus a	Addition
d. $12 - b$	12 minus b	Subtraction

Example 2 *Evaluate the Variable Expression*

Evaluate the variable expression when $x = 5$.

Solution

Expression	Substitute	Simplify
a. $5x$	$= 5(\underline{5})$	$= \underline{25}$
b. $\frac{40}{x}$	$= \frac{40}{\boxed{5}}$	$= \underline{8}$

✓ *Checkpoint* **Evaluate the variable expression when $x = 4$.**

1. $10x$	2. $\frac{20}{x}$	3. $10 - x$	4. $x + 8$
40	5	6	12

Example 3 *Evaluate rt to Find Distance*

Find the distance d traveled in 30 minutes by a remote control car going an average speed of 10 miles per hour. Use the formula: distance equals rate r multiplied by time t.

Solution

To evaluate a variable expression, you write the expression, substitute a number for each variable, and simplify.

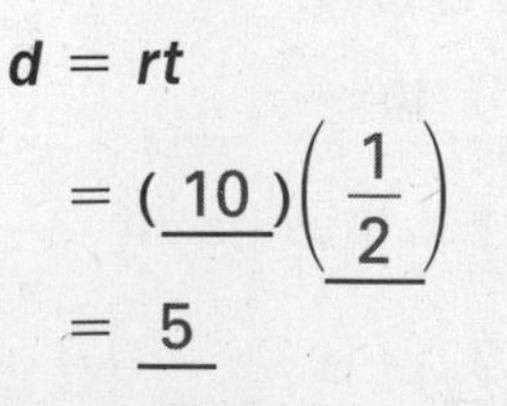

$d = rt$ **Write formula.**

$= (\underline{10})\left(\underline{\frac{1}{2}}\right)$ **Substitute for *r* and *t*.**

$= \underline{5}$ **Simplify.**

Answer The distance traveled by the remote control car was $\underline{5}$ miles.

 ***Checkpoint*** **Complete the following exercise.**

5. Using a variable expression, find the distance traveled in 20 minutes by a remote control car moving at an average speed of 9 miles per hour.

3 miles

Example 4 ***Find the Perimeter***

The perimeter P of a triangle is equal to the sum of the lengths of its sides:

$$P = a + b + c$$

$a = 6$ m, $b = 11$ m, $c = 13$ m

Find the perimeter of the triangle in meters.

Solution

1. Write the formula. $P = a + b + c$

2. Substitute 6 for a, 11 for b, and 13 for c. $= \underline{6} + \underline{11} + \underline{13}$

3. Simplify the formula. $= \underline{30}$

Answer The triangle has a perimeter of <u>30</u> meters.

Checkpoint **Complete the following exercise.**

6. Find the perimeter of a triangle with sides 7 centimeters, 12 centimeters, and 17 centimeters long.

36 cm

1.2 Exponents and Powers

Goal Evaluate a power.

VOCABULARY

Power A power is an expression of the form a^b or the value of such an expression. For example, 2^4 is a power, and because $2^4 = 16$, 16 is the fourth power of 2.

Exponent In exponential notation, the number of times the base is used as a factor is the exponent. For example, 6 is the exponent in the expression 4^6.

Base In exponential notation, the number or variable that undergoes repeated multiplication is the base. For example, 4 is the base in the expression 4^6.

Grouping symbols Symbols such as parentheses ( ) and brackets [ ] that indicate the order in which operations should be performed are grouping symbols. Operations within the innermost set of grouping symbols are done first.

Example 1 ***Read and Write Powers***

Note that x^1 is customarily written as x with the exponent omitted.

Exponential Form	Words	Meaning
a. 15^1	fifteen to the first power	15
b. 5^2	five to the second power, or five squared	$5 \cdot 5$
c. 6^3	six to the third power, or six cubed	$6 \cdot 6 \cdot 6$
d. 9^5	nine to the fifth power	$9 \cdot 9 \cdot 9 \cdot 9 \cdot 9$

✓ ***Checkpoint*** **Write the expression in exponential form.**

1. 4 cubed	**2.** z to the ninth power	**3.** a to the fifth power
4^3	z^9	a^5

Example 2 *Evaluate the Power*

Evaluate x^5 when $x = 2$.

1. Substitute 2 for x. $x^5 = 2^5$

2. Write out the factors. $= 2 \cdot 2 \cdot 2 \cdot 2 \cdot 2$

3. Multiply the factors. $= 32$

Answer The value of the expression is 32.

Example 3 *Evaluate Exponential Expressions*

Evaluate the variable expression when $a = 5$ and $b = 3$.

a. $(a - b)^4 = (5 - 3)^4$ **Substitute for *a* and *b*.**

$= 2^4$ **Subtract within parentheses.**

$= 2 \cdot 2 \cdot 2 \cdot 2$ **Write factors.**

$= 16$ **Multiply.**

b. $(a^2) - (b^2) = (5^2) - (3^2)$ **Substitute for *a* and *b*.**

$= 25 - 9$ **Evaluate powers.**

$= 16$ **Subtract.**

Checkpoint **Evaluate the variable expression when $a = 7$ and $b = 3$.**

4. $(a^2) + b$	**5.** $(a + b)^2$	**6.** $(b^2) - a$
52	100	2
7. $(a - b)^3$	**8.** $(a^2) + (b^3)$	**9.** $(a^2) - (b^2)$
64	76	40

Example 4 *Exponents and Grouping Symbols*

Notice that in part (a) of Example 4, the exponent applies to x, while in part (b) the exponent applies to $4x$.

Evaluate the variable expression when $x = 6$.

a. $4x^2 = 4(\underline{6}^2)$ **Substitute for x.**

$= 4(\underline{36})$ **Evaluate power.**

$= \underline{144}$ **Multiply.**

b. $(4x)^2 = (4 \cdot \underline{6})^2$ **Substitute for x.**

$= \underline{24}^2$ **Multiply within parentheses.**

$= \underline{576}$ **Evaluate power.**

Example 5 *Find the Volume of the Aquarium*

Aquarium An aquarium has the shape of a cube. Each edge x is 5 feet long. Find the volume in cubic feet.

Solution

$V = x^3$ **Write formula for volume of a cube.**

$= \underline{5}^3$ **Substitute for x.**

$= \underline{125}$ **Evaluate power.**

Answer The volume of the aquarium is $\underline{125}$ cubic feet.

✔ *Checkpoint* **Complete the following exercise.**

10. Use the formula for the area of a square to find the area of one side of the aquarium in Example 5. Express your answer in square feet.

25 ft^2

1.3 Order of Operations

Goal Use the established order of operations.

VOCABULARY

Order of Operations The order of operations are the rules for evaluating an expression involving more than one operation.

Left-to-Right Rule When operations have the same priority, you perform them in order from left to right.

ORDER OF OPERATIONS

Step 1 First do operations that occur within grouping symbols.

Step 2 Then evaluate powers.

Step 3 Then do multiplications and divisions from left to right.

Step 4 Finally, do additions and subtractions from left to right.

Example 1 ***Evaluate Without Grouping Symbols***

Evaluate the expression $4x^2 + 3$ when $x = 3$. Use the order of operations.

$4x^2 + 3 = 4 \cdot 3^2 + 3$	**Substitute for *x*.**
$= 4 \cdot 9 + 3$	**Evaluate power.**
$= 36 + 3$	**Evaluate product.**
$= 39$	**Add.**

Checkpoint **Evaluate the variable expression when $x = 5$. Use the order of operations.**

1. $x^2 - 10$	2. $3x^2 + 9$	3. $43 - x^2$	4. $2x^2 + 16$
15	84	18	66

Example 2 *Using the Left-to-Right Rule*

a. $28 - 7 - 4 = (28 - 7) - 4$ Work from left to right.

$= \underline{21} - 4$ Subtract $\underline{7}$ from $\underline{28}$.

$= \underline{17}$ Subtract 4 from $\underline{21}$.

b. $15 + 9 \div 3 - 4 = 15 + (9 \div 3) - 4$ Divide first.

$= 15 + \underline{3} - 4$ Divide $\underline{9}$ by $\underline{3}$.

$= (15 + \underline{3}) - 4$ Work from left to right.

$= \underline{18} - 4$ Add 15 and $\underline{3}$.

$= \underline{14}$ Subtract 4 from $\underline{18}$.

You divide first in part (b) of Example 2 because division has a higher priority than addition and subtraction.

Example 3 *Expressions with Fraction Bars*

Evaluate the expression. Then simplify the answer.

$\frac{5 \cdot 3}{13 + 6^2 - 4} = \frac{5 \cdot 3}{13 + \boxed{36} - 4}$ Evaluate power.

$= \frac{\boxed{15}}{13 + \boxed{36} - 4}$ Simplify the numerator.

$= \frac{\boxed{15}}{\boxed{49} - 4}$ Work from left to right.

$= \frac{15}{45}$ Subtract.

$= \frac{1}{3}$ Simplify.

Checkpoint Evaluate the expression when $x = 3$.

5. $1 + x^3 - 12$	6. $5x^2 - 13 + 6$	7. $\frac{3x^2}{5 + x^2 - 11}$
16	38	9

Example 4 ***Use a Calculator***

Enter the following in your calculator. Does the calculator display 5 or 9?

6 [+] 12 [÷] 3 [−] 1 [ENTER]

Solution

a. If your calculator uses the order of operations, it will display 9.

$$6 + 12 \div 3 - 1 = 6 + (12 \div 3) - 1$$
$$= 6 + \underline{4} - 1$$
$$= \underline{10} - 1$$
$$= \underline{9}$$

b. If your calculator displays 5, it performs the operations as they are entered.

$$[(6 + 12) \div 3] - 1 = (\underline{18} \div 3) - 1$$
$$= \underline{6} - 1$$
$$= \underline{5}$$

✓ *Checkpoint* **In Exercises 8 and 9, two calculators were used to evaluate the expression. Which calculator used the established order of operations?**

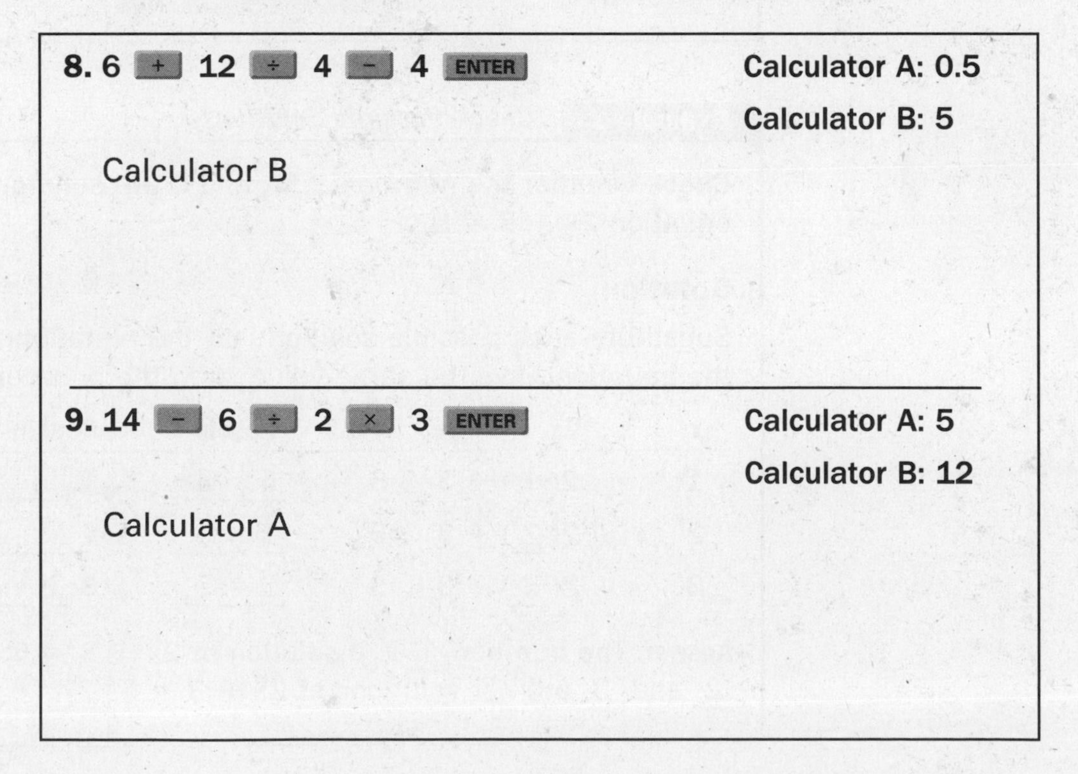

8. 6 [+] 12 [÷] 4 [−] 4 [ENTER]

Calculator A: 0.5

Calculator B: 5

Calculator B

9. 14 [−] 6 [÷] 2 [×] 3 [ENTER]

Calculator A: 5

Calculator B: 12

Calculator A

1.4 Equations and Inequalities

Goal Check solutions of equations and inequalities.

VOCABULARY

Equation An equation is a statement formed by placing an equal sign between two expressions.

Solution A number that, when substituted for the variable in an equation or inequality, results in a true statement

Inequality A statement formed by placing an inequality symbol, such as < or ≤, between two expressions

Example 1 ***Check Possible Solutions***

Check whether the numbers 1, 2, and 3 are solutions of the equation $2x + 3 = 5$.

Solution

Substitute each possible solution into the equation. If both sides of the equation have the same value, then the number is a solution.

x	$2x + 3 = 5$	Result	Conclusion
1	$2(1) + 3 \stackrel{?}{=} 5$	$5 = 5$	1 is a solution.
2	$2(2) + 3 \stackrel{?}{=} 5$	$7 \neq 5$	2 is not a solution.
3	$2(3) + 3 \stackrel{?}{=} 5$	$9 \neq 5$	3 is not a solution.

Answer The number 1 is a solution of $2x + 3 = 5$. The numbers 2 and 3 are not solutions of $2x + 3 = 5$.

Example 2 *Solve Equations with Mental Math*

To solve equations with mental math, think of the equation as a question.

Solving the equation is finding all the solutions of an equation.

Equation	Question	Solution
$3x = 12$	3 times what number gives 12?	$3 \cdot 4 = 12$, so $x = 4$
$x - 4 = 9$	What number minus 4 equals 9?	$13 - 4 = 9$, so $x = 13$
$9 = x + 6$	9 equals what number plus 6?	$9 = 3 + 6$, so $x = 3$

Example 3 *Use Mental Math to Solve a Real-Life Equation*

School Supplies You are buying school supplies. A box of pencils costs $3.08, a package of pens costs $3.45, and a single-subject notebook costs $2.79. The tax is included. You have $8. About how much more money do you need?

Solution

You can ask the question: The total cost equals 8 plus what number of dollars? Let x represent any additional money you may need. Use rounding to estimate the total cost.

$$3 + 3.5 + 3 = 8 + x$$

$$9.5 = 8 + x$$

Answer Because the total cost of the supplies is approximately 9.5 or $9.50, you can see that you need about $1.50 more to purchase all the supplies.

✔ ***Checkpoint*** **Use the information in Example 3 to complete the following exercise.**

1. You decide to buy a 5-subject notebook for $5.89 instead of the single-subject notebook for $2.79. About how much more money will you need to buy the 5-subject notebook and the rest of the supplies?

$4.50

Example 4 *Check Solutions of Inequalities*

The "wide end" of the inequality symbol faces the greater number.

Decide whether 3 is a solution of the inequality.

a. $x + 5 < 9$ **b.** $3x + 2 \leq 11$ **c.** $x - 1 > 4$

Solution

Inequality	Substitute	Result	Conclusion
a. $x + 5 < 9$	$\underline{3} + 5 \overset{?}{<} 9$	$\underline{8 < 9}$	3 $\underline{\text{is}}$ a solution.
b. $3x + 2 \leq 11$	$3(\underline{3}) + 2 \overset{?}{\leq} 11$	$\underline{11 \leq 11}$	3 $\underline{\text{is}}$ a solution.
c. $x - 1 > 4$	$\underline{3} - 1 \overset{?}{>} 4$	$\underline{2 \not> 4}$	3 $\underline{\text{is not}}$ a solution.

Checkpoint **Decide whether the given number is a solution of the inequality.**

2. $35 - 7c > 20$; 2 solution	**3.** $5n - 7 < 23$; 6 not a solution
4. $x^2 + 6 \geq 55$; 7 solution	**5.** $\frac{63}{y} \leq 21$; 2 not a solution

1.5 Translating Words into Mathematical Symbols

Goal Translate words into mathematical symbols.

VOCABULARY

Translate Use mathematical symbols to represent written statements or real-life situations.

Phrase A phrase is a mathematical expression in words.

Sentence A mathematical sentence with an equal sign is an equation.

Example 1 ***Translate Addition and Subtraction Phrases***

Write the phrase as a variable expression. Let x represent the number.

Phrase	Translation
A number *plus* 9	$x + 9$
The *sum* of 12 and a number	$12 + x$
A number *increased* by 2	$x + 2$
19 *more than* a number	$x + 19$
A number *decreased* by 17	$x - 17$
The *difference* between 12 and a number	$12 - x$
13 *minus* a number	$13 - x$
1 *less than* a number	$x - 1$

Order is important for subtraction. "4 less than a number" means $y - 4$, not $4 - y$.

Checkpoint **Write the phrase as a variable expression. Let x represent the number.**

1. 17 minus a number	2. A number increased by 5
$17 - x$	$x + 5$

Example 2 *Translate Multiplication and Division Phrases*

Write the phrase as a variable expression. Let *n* represent the number.

Phrase	Translation
15 *times* a number	$15n$
A number *multiplied* by 7	$7n$
The *product* of 8 and a number	$8n$
The *quotient* of a number and 9	$\frac{n}{9}$
One third *of* a number	$\frac{1}{3}n$
12 *divided* by a number	$\frac{12}{n}$

Example 3 *Translate Sentences*

Phrases are translated into variable expressions. Sentences are translated into equations or inequalities.

Write the sentence as an equation or an inequality.

Sentence	Translation
A number x decreased by 4 is less than 18.	$x - 4 < 18$
The product of 11 and a number x is 33.	$11x = 33$
The sum of 7 and a number x is greater than 53.	$7 + x > 53$

Checkpoint **Write each phrase as a variable expression. Let *x* represent the number.**

3. 19 times a number	**4.** A number divided by 14
$19x$	$\frac{x}{14}$

Write the sentence as an equation or an inequality.

5. The quotient of 40 and a number x is 8.	**6.** The difference between 17 and a number x is 5.
$\frac{40}{x} = 8$	$17 - x = 5$

Example 4 *Write and Solve an Equation*

Translate into mathematical symbols: "The product of 9 and a number is 72." Let x represent the number. Use mental math to solve the equation. Then check your solution.

Solution

The equation is $9x = 72$. Using mental math, you can find that the solution is $x = 8$.

Check		
	$9x = 72$	**Write original equation.**
	$9(8) \stackrel{?}{=} 72$	**Substitute for x.**
	$72 = 72$	**Solution checks.**

Example 5 *Translate and Solve a Real-Life Problem*

You and your friends go to a video store to buy DVDs on sale for $13 each (including tax). Together you spent $78. Use mental math to solve the equation for how many DVDs you and your friends bought.

Solution

Let x represent the number of DVDs you and your friends bought.

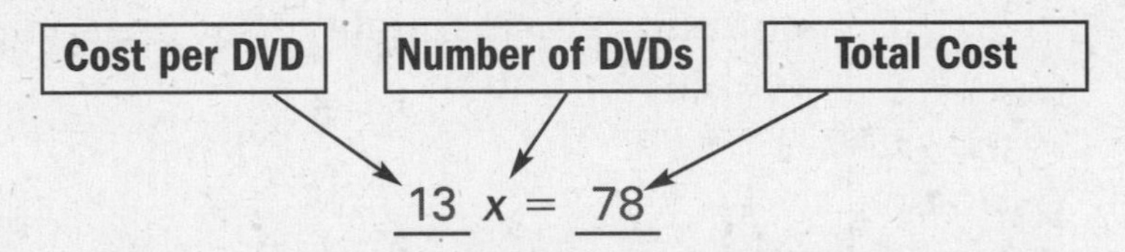

$$13\,x = 78$$

Ask what number times 13 equals 78. Use mental math to find $x = 6$.

Answer You and your friends bought 6 DVDs.

✓ *Checkpoint* **Complete the following exercise.**

7. The science club is selling magazine subscriptions at $15 each. How many subscriptions does the club have to sell to raise $300?

   20 subscriptions

1.6 A Problem Solving Plan Using Models

Goal Model and solve real-life problems.

VOCABULARY

Modeling Representing real-life situations by means of equations or inequalities is called modeling.

Verbal Model A verbal model is an expression that uses words to describe a real-life situation.

Algebraic Model An algebraic model is an expression, equation, or inequality that uses variables to represent a real-life situation.

A PROBLEM SOLVING PLAN USING MODELS

Verbal Model — Ask yourself what you need to know to solve the problem. Then write a verbal model that will give you what you need to know.

↓

Labels — Assign labels to each part of your verbal model.

↓

Algebraic Model — Use the labels to write an algebraic model based on your verbal model.

↓

Solve — Solve the algebraic model and answer the original question.

↓

Check — Check that your answer is reasonable.

Example 1 *Write an Algebraic Model*

You go to a music store to buy music CDs on sale for $8 each. You spend $42.40, including $2.40 for sales tax. Use modeling to find the number of CDs you bought.

> Be sure you understand the problem before you write a model. For example, notice that the tax is added after the cost of the CDs is figured.

Solution

Verbal Model [Cost per CD] • [Number of CDs] = [Total Cost] − [Tax]

Labels		
	Cost per CD = 8	**(dollars)**
	Number of CDs = d	**(CDs)**
	Total cost = 42.40	**(dollars)**
	Tax = 2.40	**(dollars)**

Algebraic Model		
	$8d = 42.40 - 2.40$	**Write algebraic model.**
	$8d = 40$	**Subtract.**
	$d = 5$	**Solve using mental math.**

Answer You bought 5 music CDs.

Checkpoint **Complete the following exercise.**

1. You go to a music store to buy music CDs for $13.50 each. You spend a total of $43.74, including $3.24 for sales tax. Use modeling to find the number of CDs you bought.

3 CDs

Example 2 *Write an Algebraic Model*

You are racing in a bicycle marathon that is 75 miles long. Your average speed is 15 miles per hour. After 3 hours, you have cycled 45 miles.

a. If you maintain your average speed, how long will it take you to finish the last 30 miles of the marathon?

b. At your average speed, is it reasonable to expect that you can finish the entire marathon in 4.5 hours?

Solution

a. Use the formula (rate)(time) = (distance) to write a verbal model.

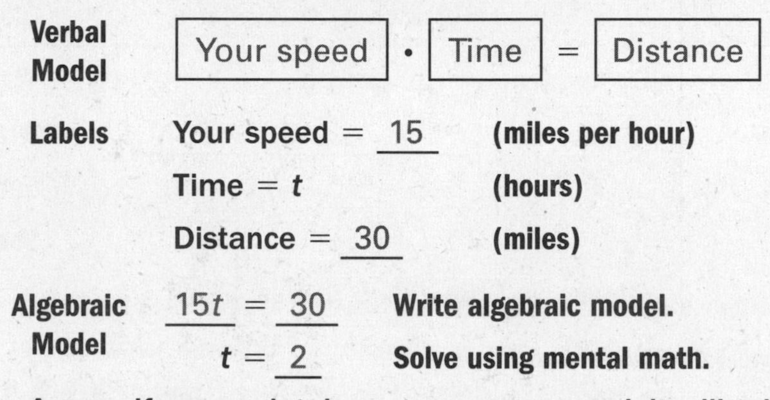

Algebraic Model $15t = 30$ **Write algebraic model.**

$t = 2$ **Solve using mental math.**

Answer If you maintain your average speed, it will take you 2 hours to finish the marathon.

b. At your average speed, it is not reasonable to expect that you can finish the entire marathon in 4.5 hours.

✔ ***Checkpoint*** **Complete the following exercise.**

2. The math club is selling flower baskets for $8 each to raise money for a trip. How many baskets of flowers does the club have to sell to raise $640? Use the problem solving plan to answer the question.

80 flower baskets

1.7 Tables and Graphs

Goal Organize data using a table or graph.

VOCABULARY

Data Data are information, facts, or numbers used to describe something.

Bar graph A bar graph is a graph that represents a collection of data by using horizontal or vertical bars whose lengths allow the data to be compared.

Line graph A line graph is a graph that uses line segments to connect data points. Line graphs are especially useful for showing changes in data over time.

Example 1 ***Organize Data in a Table***

The table gives the media consumer spending (dollars per person per year) in the United States. Make a table showing the total media spending for daily newspapers, consumer magazines, and consumer books. In which year did Americans spend the most?

Top Categories for Media Consumer Spending

Year	1996	1997	1998	1999	2000
Cable and satellite TV	\$138.96	\$153.11	\$165.56	\$179.89	\$192.82
Daily newspaper	\$52.84	\$52.81	\$53.30	\$53.65	\$53.32
Consumer magazines	\$39.51	\$40.33	\$40.57	\$40.30	\$39.50
Consumer books	\$72.68	\$72.26	\$75.62	\$80.43	\$77.64

Solution **To make the table, add the amounts for daily newspapers, consumer magazines, and consumer books for each year. From the table, you can see that the greatest consumer spending was in** 1999 **.**

Year	1996	1997	1998	1999	2000
Total	\$165.03	\$165.40	\$169.49	\$174.38	\$170.46

Example 2 *Interpret a Bar Graph*

The bar graph shows the amount Americans spent for cable and satellite television in a given year. It appears that Americans spent twice as much in 2000 as compared with 1996. Explain why the graph could be misleading.

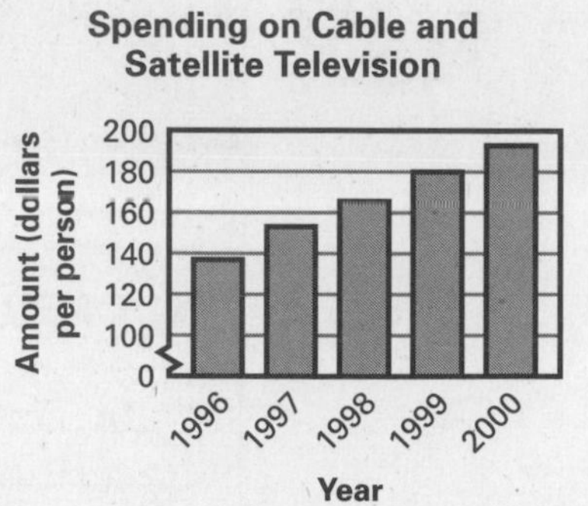

Solution

The bar graph could be misleading because the vertical scale is not consistent. The zigzag line shows a break where part of the scale is not shown. Because of the break, the first tick mark on the vertical scale represents $100.00 spent on cable or satellite television. The other tick marks on the scale each represent $20.00 spent on cable or satellite television.

Checkpoint **Complete the following exercise.**

1. Draw a new bar graph for Example 2 that would not be misleading.

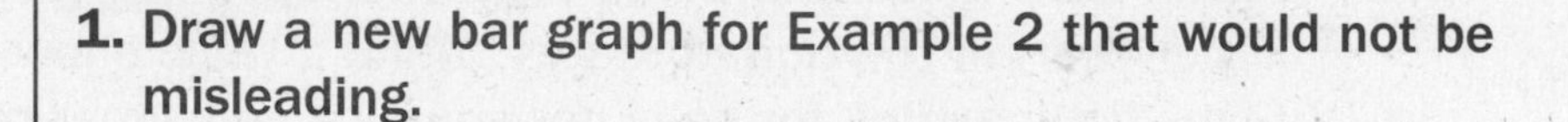

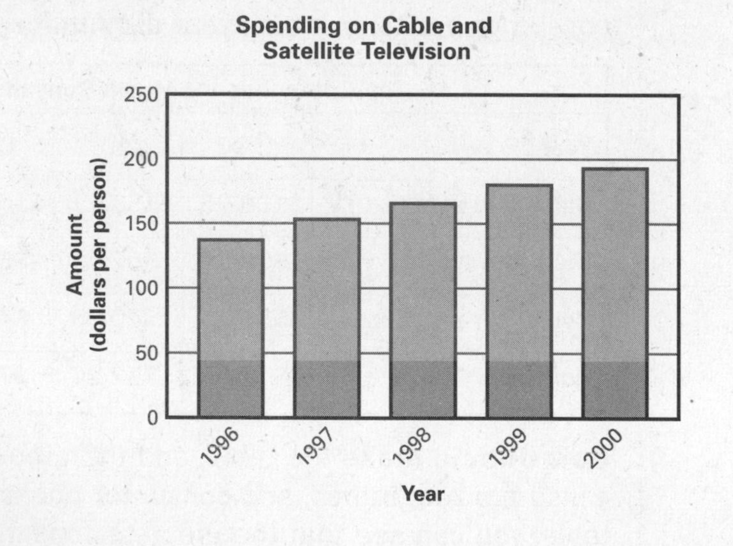

Example 3 *Make and Interpret a Line Graph*

From 1994 to 2001, the average amount spent per student in the United States in public elementary and secondary schools is given in the table. Draw a line graph of the data.

Average Amount Spent per Student								
Year	1994	1995	1996	1997	1998	1999	2000	2001
Amount	$5996	$6208	$6443	$6764	$7142	$7533	$7885	$8194

Solution

Each unit on the vertical axis represents 1000 dollars. Each unit on the horizontal axis represents 1 year, starting with 1994.

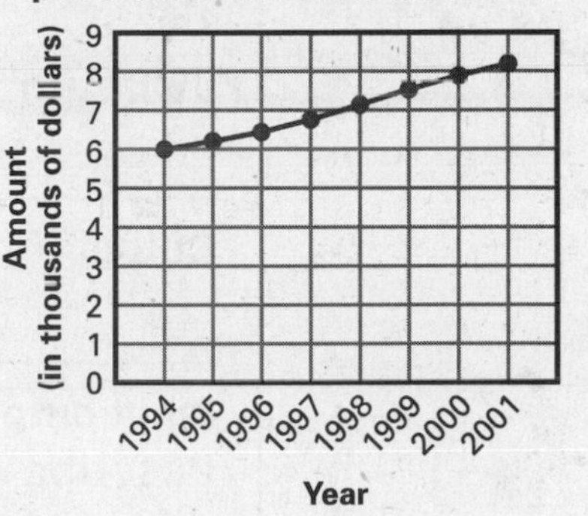

Checkpoint Complete the following exercise.

2. Using the data from Example 1, make a table showing the total media spending (dollars per person per year) for cable and satellite TV, daily newspaper, consumer magazines, and consumer books. Draw a line graph from the table.

Media Consumer Spending (dollars per person per year)					
Year	1996	1997	1998	1999	2000
Total	$303.99	$318.51	$335.05	$354.27	$363.28

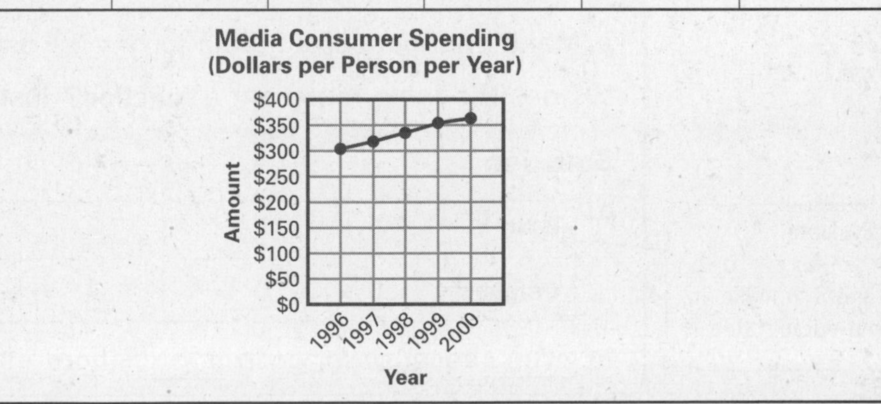

An Introduction to Functions

Goal Use four different ways to represent functions.

VOCABULARY

Function A function is a rule that establishes a relationship between two quantities, called the input and output. There is exactly one output for each input.

Input An input is a value in the domain of a function.

Output An output is a value in the range of a function.

Input-output table An input-output table is a table used to describe a function by listing the outputs for several different inputs.

Domain The domain of a function is the collection of all input values of a function.

Range The range of a function is the collection of all output values of a function.

Example 1 ***Make an Input-Output Table***

a. Make an input-output table for $y = x^2$ using $x = 0, 1, 2,$ and 3.

b. Does the table represent a function? Justify your answer.

Solution

Evaluate $y = x^2$ for $x = 0, 1, 2,$ and 3 to make an input-output table.

a.

Input x	0	1	2	3
Output y	0	1	4	9

b. Yes, because for each input there is exactly one output.

Checkpoint Complete the following exercise.

1. Make an input-output table for $y = 2x - 1$ using $x = 1, 2, 3,$ and 4. Does the table represent a function? Justify your answer.

Input x	1	2	3	4
Output y	1	3	5	7

Yes, because for each input there is exactly one output.

Example 2 *Use a Table to Graph a Function*

Airplane An airplane is at an altitude of 34,000 feet. The pilot raises the airplane at a rate of 200 feet per minute for 10 minutes. The altitude h in feet after the airplane has risen for t minutes is given by $h = 34{,}000 + 200t$, where $t \geq 0$ and $t \leq 10$.

The domain is the collection of all input values of a function; the range is the collection of all output values of a function.

a. Make an input-output table for the function.

b. Draw a graph that represents the function.

Solution

a.

Input t	0	2	4	6	8	10
Output h	34,000	34,400	34,800	35,200	35,600	36,000

b. Let the horizontal axis represent the input t (in minutes). Label the axis from 0 to 10. Let the vertical axis represent the output h (in feet). Label the axis from 0 to 40,000.

Plot and connect the data points given in the table.

The graph shows that as the time increases, the height of the airplane increases.

The graph represents the function $h = 34{,}000 + 200t$, where $t \geq 0$ and $t \leq 10$.

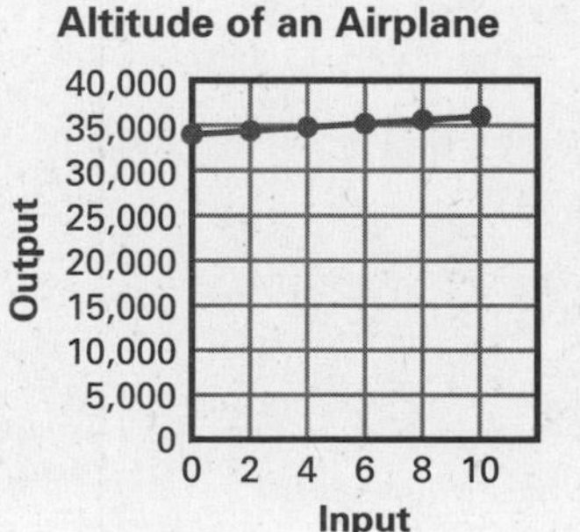

Example 3 *Write an Equation to Represent a Function*

Auto Repair To fix a car, a mechanic tells the owner that the parts will cost \$270. The cost for the mechanic to fix the car is \$35 per hour. Write an equation to represent the total cost of the repair C as a function of the hours h that it takes to fix the car.

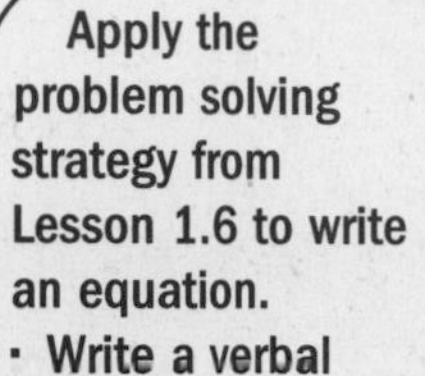

Apply the problem solving strategy from Lesson 1.6 to write an equation.
- Write a verbal model.
- Assign labels.
- Write an algebraic model.

Solution

Verbal Model

Total cost	=	Cost of parts	+	Cost per hour	•	Hours

Labels

Total cost $= C$ **(dollars)**

Cost of parts = 270 **(dollars)**

Cost per hour = 35 **(dollars)**

Hours $= h$ **(hours)**

Algebraic Model $C = 270 + 35h$

Checkpoint Complete the following exercise.

2. You are buying fabric to make costumes for the school play. The cost for the fabric is \$6.40 per yard. Write an equation to represent the total cost of the fabric C as a function of the yards of fabric y that you buy.

$C = 6.40y$

Words to Review

Give an example of the vocabulary word.

Variable expression
$5x$

Verbal model
Distance = Rate • Time

Power
6^4

Exponent
In the expression 6^4, 4 is the exponent.

Base of a power
In the expression 6^4, 6 is the base.

Equation
$x + 3 = 5$

Solution of an equation
The solution of the equation $x + 3 = 5$ is 2.

Inequality
$x + 3 < 5$

Function
$y = 2x$

Input-output table

Input	0	1	2
Output	0	2	4

Domain
For this input-output table, the domain is 0, 1, and 2.

Input	0	1	2
Output	0	2	4

Range
For this input-output table, the range is 0, 2, and 4.

Input	0	1	2
Output	0	2	4

Review your notes and Chapter 1 by using the Chapter Review on pages 55–58 of your textbook.

2.1 The Real Number Line

Goal Graph, compare, and order real numbers.

VOCABULARY

Real numbers Real numbers are the set of numbers consisting of the positive numbers, the negative numbers, and zero.

Real number line A real number line is a line whose points correspond to the real numbers.

Positive numbers Positive numbers are any of the numbers greater than zero.

Negative numbers Negative numbers are any of the numbers less than zero.

Integers Any of the numbers ..., −3, −2, −1, 0, 1, 2, 3, ... are examples of integers.

Whole numbers The whole numbers are the set of numbers consisting of the positive integers and zero.

Graph of a number The graph of a number is the point on a number line that corresponds to the number.

In Example 1, −5 is read as "negative 5," −1 is read as "negative 1," and 2 is read as "two" or as "positive two."

Example 1 ***Graph Integers***

Graph −5, −1, and 2 on a number line.

Solution

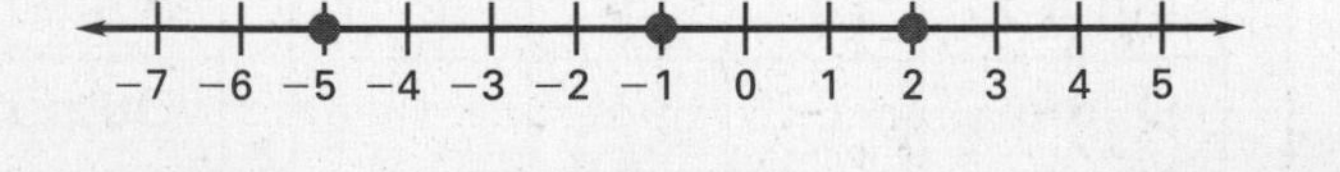

−5 is negative **, so it is plotted** 5 units **to the** left **of zero.**

−1 is negative **, so it is plotted** 1 unit **to the** left **of zero.**

2 is positive **, so it is plotted** 2 units **to the** right **of zero.**

Example 2 *Compare Integers*

Graph −3 and −7 on a number line. Then write two inequalities that compare the two numbers.

Solution

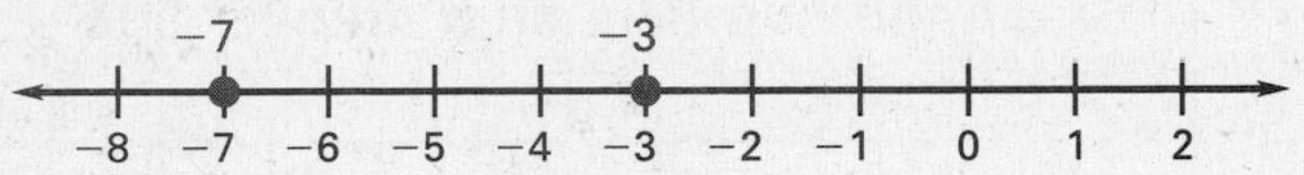

On the graph, −7 is to the left of −3, so −7 is less than −3.

On the graph, −3 is to the right of −7, so −3 is greater than −7.

Answer $-7 < -3$ and $-3 > -7$

Checkpoint Complete the following exercise.

1. Graph −8 and 2 on a number line. Then write two inequalities that compare the numbers.

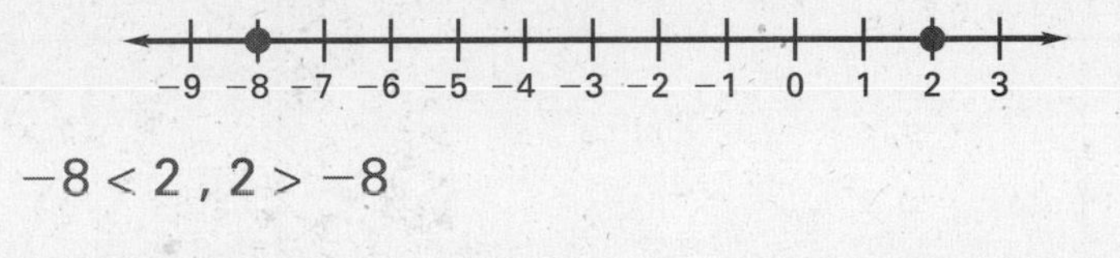

Example 3 *Graph Real Numbers*

Graph $-\frac{2}{3}$ and 0.4 on a number line.

Solution

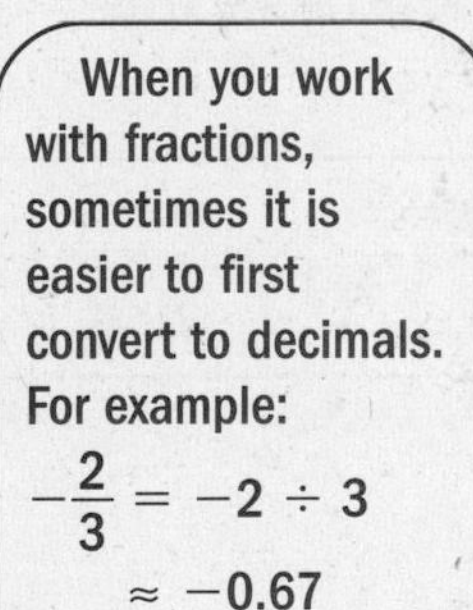

Because $-\frac{2}{3}$ and 0.4 are not integers, use a number line that has scale marks in smaller units.

On the graph, $-\frac{2}{3}$ is a negative number, which is approximately −0.67. Plot the point 0.67 units to the left of zero.

On the graph, 0.4 is a positive number, so plot the point 0.4 units to the right of zero.

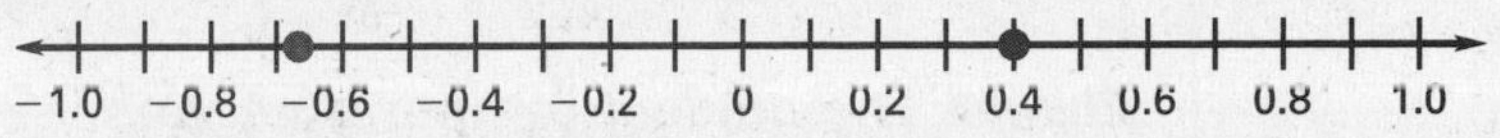

Example 4 *Order Real Numbers*

Write the numbers -3, 4.5, $-\frac{4}{5}$, 1.8, $\frac{3}{2}$, and -4 in increasing order.

Solution

Graph the numbers on a number line.

−5 −4 −3 −2 −1 0 1 2 3 4 5

Answer From the graph, you can see that the order is

$-4, -3, -\frac{4}{5}, \frac{3}{2}, 1.8, 4.5$.

Checkpoint Write the numbers in increasing order.

2. $-\frac{8}{9}$, 0.9, −1.8, −2, 3, $\frac{1}{9}$	3. 4, −1.3, $\frac{1}{2}$, −4.2, 0, $\frac{13}{5}$
$-2, -1.8, -\frac{8}{9}, \frac{1}{9}, 0.9, 3$	$-4.2, -1.3, 0, \frac{1}{2}, \frac{13}{5}, 4$

Example 5 *Compare Real Numbers*

Low Temperature The table shows the lowest monthly temperatures in Des Moines, Iowa for the first six months of the year. Which temperature was the coldest?

January	February	March	April	May	June
−24°F	−26°F	−22°F	9°F	30°F	38°F

Solution

First graph the temperatures on a number line.

−40 −35 −30 −25 −20 −15 −10 −5 0 5 10 15 20 25 30 35 40

Answer The coldest temperature was −26°F in February.

Absolute Value

Goal Find the opposite and the absolute value of a number.

VOCABULARY

Opposite Two numbers that are the same distance from 0 on a number line but on opposite sides of 0 are opposites.

Absolute value The absolute value of a number is its distance from zero on a number line.

Counterexample A counterexample is a single example used to show that a given statement is false.

Example 1 ***Find the Opposite of a Number***

Use a number line to find the opposite of 6.

−7 −6 −5 −4 −3 −2 −1 0 1 2 3 4 5 6 7

On the number line, 6 is 6 **units to the** right **of 0. The opposite of 6 is** 6 **units to the** left **of 0. So the opposite of 6 is** −6**.**

THE ABSOLUTE VALUE OF A NUMBER

- **If a is a positive number, then $|a| =$** a **.**
  Example: $|9| = 9$
- **If a is zero, then $|a| =$** 0 **.**
  Example: $|0| = 0$
- **If a is a negative number, then $|a| =$** $-a$ **.**
  Example: $|-9| = -(-9) = 9$

The expression $-a$ can be read as "negative a" or as "the opposite of a."

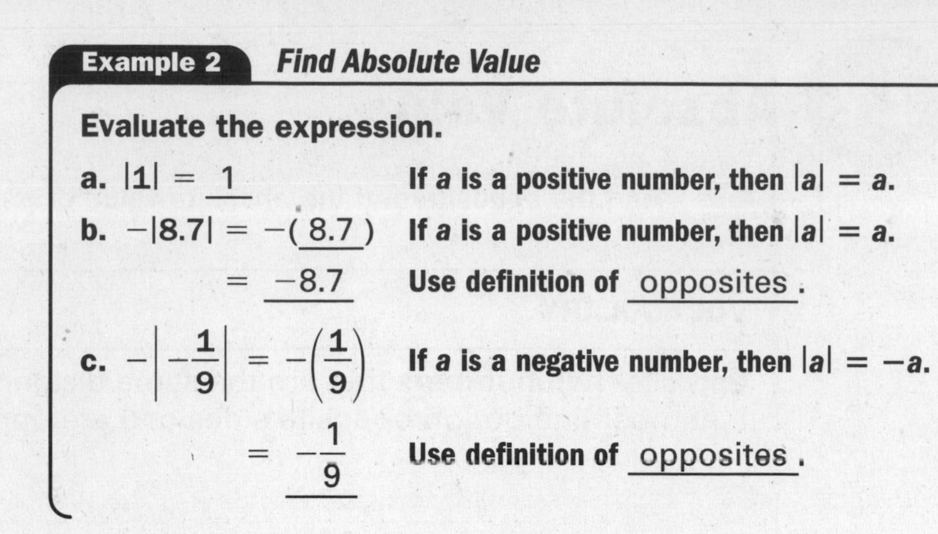

Example 2 ***Find Absolute Value***

Evaluate the expression.

a. $|1| = \underline{1}$ — **If *a* is a positive number, then $|a| = a$.**

b. $-|8.7| = -(\underline{8.7})$ — **If *a* is a positive number, then $|a| = a$.**

$= \underline{-8.7}$ — **Use definition of** opposites **.**

c. $-\left|-\frac{1}{9}\right| = -\left(\frac{1}{9}\right)$ — **If *a* is a negative number, then $|a| = -a$.**

$= \underline{-\frac{1}{9}}$ — **Use definition of** opposites **.**

Example 3 ***Solve an Absolute Value Equation***

Use mental math to solve the equation.

a. $|x| = 2$ **b.** $|x| = -2$

Solution

a. Both 2 and −2 are 2 units from 0, so there are two solutions: 2 and −2 .

b. Because distance is never negative, the absolute value of a number is never negative, so there is no solution .

Checkpoint **Evaluate the expression.**

1. $\lvert -10 \rvert$	**2.** $-\lvert 6.1 \rvert$	**3.** $-\lvert -3 \rvert$	**4.** $\lvert 3.8 \rvert$
10	−6.1	−3	3.8

Use mental math to solve the equation. If there is no solution, write *no solution*.

5. $\lvert x \rvert = \frac{3}{5}$	**6.** $\lvert x \rvert = 4.9$	**7.** $\lvert x \rvert = -\frac{4}{11}$	**8.** $\lvert x \rvert = 5$
$x = -\frac{3}{5}, \frac{3}{5}$	$x = -4.9, 4.9$	no solution	$x = -5, 5$

Example 4 *Find Velocity and Speed*

A rock falls from a bridge at a rate of 9.8 meters per second. What are its velocity and speed?

> *Velocity* indicates both speed and direction (up is positive and down is negative). The *speed* of an object is the absolute value of its velocity.

Solution

Velocity = -9.8 meters per second **Motion is** downward **.**

Speed = $|-9.8|$ **Speed is never negative.**

= 9.8 meters per second

Example 5 *Using a Counterexample*

Determine whether the statement is *true* or *false*. If it is false, give a counterexample.

a. The absolute value of a negative number is *never* positive.

b. The expression $-|a|$ is *always* less than a.

Solution

a. False. Counterexample: The absolute value of -1 is 1, which is positive.

b. False. Counterexample: If $a = -3$, then $-|a| = -3$, which is not less than -3.

✓ *Checkpoint* Complete the following exercises.

9. A hotel elevator rises from the 1st floor to the 29th floor at a rate of 63 feet per minute. What are the elevator's velocity and speed?

63 feet per minute; 63 feet per minute

10. Determine whether the following statement is *true* or *false*. If it is false, give a counterexample.

The expression $-a$ is always negative.

False. Counterexample: if $a = -4$, then $-a = -(-4) = 4$.

2.3 Adding Real Numbers

Goal Add real numbers using a number line or the rules of addition.

Example 1 *Add Using a Number Line*

> You add a positive number by moving to the right on the number line. You add a negative number by moving to the left on the number line.

Use a number line to find the sum.

a. $-2 + 7$

b. $8 + (-5)$

Solution

a. Start at -2. Then move 7 units to the right.

−7 −6 −5 −4 −3 −2 −1 0 1 2 3 4 5 6 7

Answer The sum can be written as $-2 + 7 = 5$.

b. Start at 8. Then move 5 units to the left.

−4 −3 −2 −1 0 1 2 3 4 5 6 7 8 9 10

Answer The sum can be written as $8 + (-5) = 3$.

Checkpoint Use a number line to find the sum.

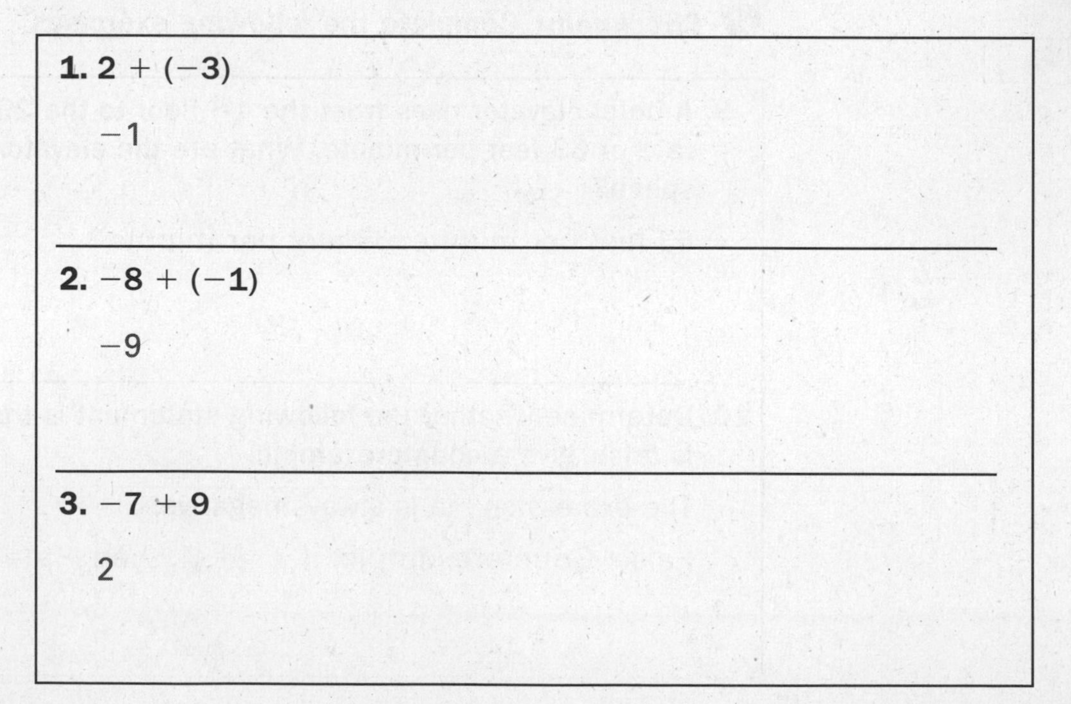

1. $2 + (-3)$

-1

2. $-8 + (-1)$

-9

3. $-7 + 9$

2

RULES OF ADDITION

To add two numbers with the *same* sign:

Step 1 Add their absolute values.

Step 2 Attach the common sign.

To add two numbers with *opposite* signs:

Step 1 Subtract the smaller absolute value from the larger one.

Step 2 Attach the sign of the number with the larger absolute value.

Example 2 ***Add Using Rules of Addition***

a. Add -9 and -12, which have the same sign.

❶ Add their absolute values. $9 + 12 = 21$

❷ Attach the common (negative) sign. $-(21) = -21$

Answer The sum can be written as $-9 + (-12) = -21$.

b. Add -13 and 6, which have opposite signs.

❶ Subtract their absolute values. $13 - 6 = 7$

❷ Attach the sign of the number with the larger absolute value. $-(7) = -7$

Answer The sum can be written as $-13 + 6 = -7$.

✔ ***Checkpoint*** **Use the rules of addition to find the sum.**

4. $17 + (-11)$	**5.** $-27 + (-14)$	**6.** $-10 + 6$
6	-41	-4

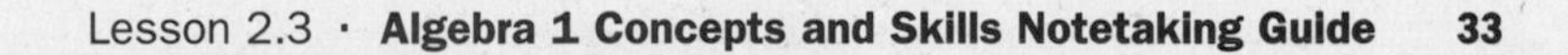

PROPERTIES OF ADDITION

Closure Property The sum of any two real numbers is a unique real number.

$a + b$ is a unique real number **Example:** $5 + 9 = 14$

Commutative Property The order in which two numbers are added does not change the sum.

$a + b =$ $b + a$ **Example:** $7 + (-12) =$ -12 $+ 7$

Associative Property The way three numbers are grouped when added does not change the sum.

$(a + b) + c =$ $a + (b + c)$ **Example:** $(1 + 6) + 8 = 1 + (6 + 8)$

Identity Property The sum of a number and 0 is the number.

$a + 0 =$ a **Example:** $12 + 0 =$ 12

Inverse Property The sum of a number and its opposite is 0.

$a + (-a) =$ 0 **Example:** $19 + (-19) =$ 0

Example 3 *Add Using Properties of Addition*

Use the properties of addition to find $-5 + 12 + 5$.

$-5 + 12 + 5 = -5 + 5 + 12$ **Use** commutative **property.**

$= (-5 + 5) + 12$ **Use** associative **property.**

$= 0 + 12$ **Use** inverse **property.**

$= 12$ **Use** identity **property.**

Checkpoint **Use the properties of addition to find the sum.**

7. $8 + (-10) + 12$	**8.** $-6 + 2 + (-4)$	**9.** $\frac{3}{4} + \left(-\frac{1}{4}\right) + \frac{1}{4}$
10	-8	$\frac{3}{4}$

2.4 Subtracting Real Numbers

Goal Subtract real numbers using the subtraction rule.

VOCABULARY

Term The terms of an expression are the parts that are added to form an expression. For example, in the expression $5 - x$, the terms are 5 and $-x$.

SUBTRACTION RULE

To subtract b from a, add the opposite of b to a.

$a - b = a + (-b)$ **Example:** $8 - 12 = 8 + (-12)$

The result is the difference of a and b.

Example 1 ***Use the Subtraction Rule***

Find the difference.

a. $-13 - 4 = -13 + (\underline{-4})$ **Add the opposite of** $\underline{-4}$.

$= \underline{-17}$ **Use the rules of addition.**

b. $-4 - (-13) = -4 + \underline{13}$ **Add the opposite of** $\underline{-13}$.

$= \underline{9}$ **Use the rules of addition.**

Example 2 ***Expressions with More than One Subtraction***

Evaluate the expression $-5 - 3 - (-1)$.

Use the left-to-right rule to evaluate the expression in Example 2.

$-5 - 3 - (-1) = -5 + (\underline{-3}) + \underline{1}$ **Add the opposites of** $\underline{3}$ **and** $\underline{-1}$.

$= \underline{-8} + \underline{1}$ **Add** $\underline{-5}$ **and** $\underline{-3}$.

$= \underline{-7}$ **Add** $\underline{-8}$ **and** $\underline{1}$.

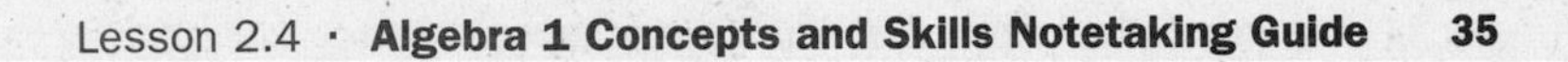

 ***Checkpoint*** **Use the subtraction rule to find the difference.**

1. $-3 - 8$ -11	2. $-12 - (-5)$ -7	3. $-7 - 4 - (-9)$ -2

Example 3 ***Evaluate a Function***

Evaluate the function $y = 2x - 3$ when $x = -2, -1, 0,$ and 1. Organize your results in a table.

Input x	Function	Output y
-2	$y = 2(\underline{-2}) - 3$	$\underline{-7}$
-1	$y = 2(\underline{-1}) - 3$	$\underline{-5}$
0	$y = 2(\underline{0}) - 3$	$\underline{-3}$
1	$y = 2(\underline{1}) - 3$	$\underline{-1}$

Checkpoint **Complete the following exercise.**

4. Evaluate the function $y = 10 - 5x$ when $x = -2, -1, 0, 1,$ and 2. Organize your results in a table.

Input x	Function	Output y
-2	$y = 10 - 5(-2)$	20
-1	$y = 10 - 5(-1)$	15
0	$y = 10 - 5(0)$	10
1	$y = 10 - 5(1)$	5
2	$y = 10 - 5(2)$	0

When an expression is written as a sum, the parts that are added are the terms of the expression.

Example 4 ***Find the Terms of an Expression***

Find the terms of the expression $-3 - 8x$.

Solution Use the subtraction rule.

$-3 - 8x = -3 + (\underline{-8x})$ **Rewrite the difference as a** $\underline{\text{sum}}$.

Answer The terms of the expression are $\underline{-3}$ and $\underline{-8x}$.

2.5 Multiplying Real Numbers

Goal Multiply real numbers using the rule for the sign of a product.

RULES FOR THE SIGN OF A PRODUCT OF NONZERO NUMBERS

- A product is negative if it has an <u>odd</u> number of negative factors.
- A product is positive if it has an <u>even</u> number of negative factors.

> In Example 1(d), be sure you understand that -3^2 is not the same as $(-3)^2$.

Example 1 ***Multiply Real Numbers***

a. $4(-6) = \underline{-24}$ **One negative factor, so product is <u>negative</u>.**

b. $7(-7)(-2) = \underline{98}$ **Two negative factors, so product is <u>positive</u>.**

c. $-1(-2)(-4) = \underline{-8}$ **Three negative factors, so product is <u>negative</u>.**

d. $-3^2(-5) = \underline{45}$ **Two negative factors, so product is <u>positive</u>.**

> Notice the similarities in the properties of multiplication and the properties of addition in Lesson 2.3.

PROPERTIES OF MULTIPLICATION

Closure Property The product of any two real numbers is a unique real number.

ab is a unique real number. **Example:** $3 \cdot 8 = 24$

Commutative Property The order in which two numbers are multiplied does not change the product.

$ab = ba$ **Example:** $(-3)8 = 8(-3)$

Associative Property The way you group three numbers when multiplying does not change the product.

$(ab)c = a(bc)$ **Example:** $(-3 \cdot 2)8 = -3(2 \cdot 8)$

PROPERTIES OF MULTIPLICATION (CONTINUED)

Identity Property The product of a number and 1 is the number.

$1 \cdot a = a$ **Example:** $1 \cdot (-3) = -3$

Property of Zero The product of a number and 0 is 0.

$0 \cdot a = 0$ **Example:** $0 \cdot (-3) = 0$

Property of Negative One The product of a number and -1 is the opposite of the number.

$-1 \cdot a = -a$ **Example:** $-1 \cdot (-3) = 3$

Example 2 ***Products with Variable Factors***

Simplify the expression.

a. $x(-y) = -xy$ — **One negative factor, so product is** negative**.**

b. $-1(-a)(-a)(-a) = (-1)(-a^3)$ — **Three negative factors, so product is** negative**.**

$= a^3$ — **Property of** negative one

c. $9(-2)(-a)^2(-b)$

$= 9(-2)(-a)(-a)(-b)$ — **Write the power as a product.**

$= 18a^2b$ — **Four negative factors, so product is** positive**.**

Checkpoint **Simplify the expression.**

1. $6(-8t)(-t)$	**2.** $(-x)(-x)(2)(-4)$	**3.** $7(-4y)(-y)(-y)^2$
$48t^2$	$-8x^2$	$28y^4$

Example 3 *Evaluate a Variable Expression*

Evaluate the expression when $x = -3$.

a. $-4x$

b. $-2x^2$

Solution

a. $-4x = -4(\underline{-3})$ **Substitute $\underline{-3}$ for x.**

$= 12$ **Two negative factors, so product is $\underline{\text{positive}}$.**

b. $-2x^2 = -2(\underline{-3})^2$ **Substitute $\underline{-3}$ for x.**

$= -2(\underline{9})$ **$\underline{\text{Evaluate}}$ power.**

$= \underline{-18}$ **One negative factor, so product is $\underline{\text{negative}}$.**

✔ *Checkpoint* Evaluate the expression when $x = -1$.

4. $-2x$ 2	**5.** $-5(-x)(-x)$ -5
6. $4(-x^2)$ -4	**7.** $-3(-x)^2$ -3

2.6 The Distributive Property

Use the distributive property.

Example 1 ***Use an Area Model***

Find the area of a rectangle whose width is 5 and whose length is $x + 3$.

Solution

You can find the area in two ways.

Area of One Rectangle

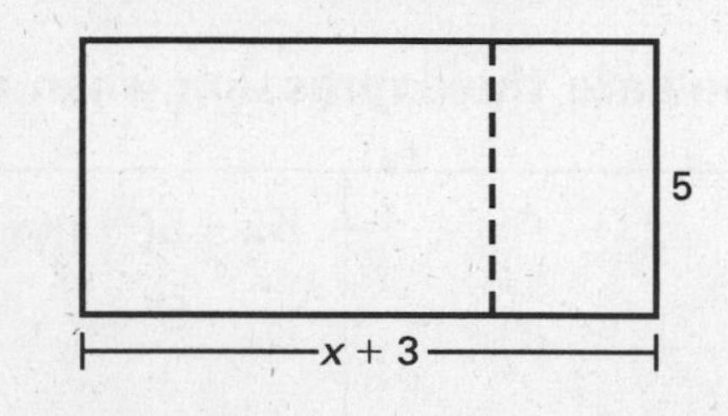

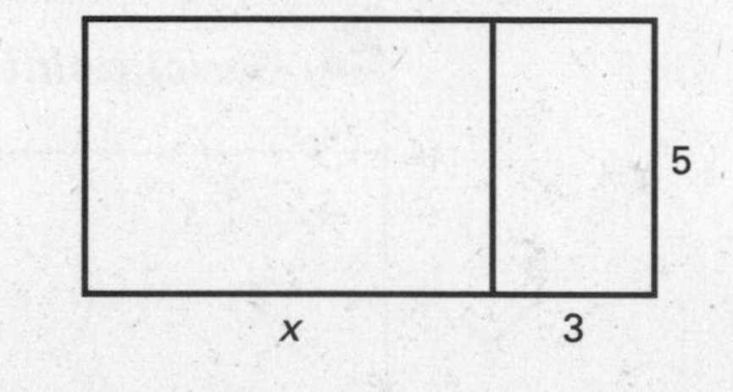

Answer Because both ways produce the same area, the following statement is true.

Area $= \underline{5}\,(\underline{x + 3}) = \underline{5}(\underline{x}) + \underline{5}(\underline{3}) = \underline{5x + 15}$

THE DISTRIBUTIVE PROPERTY

The product of a and $(b + c)$:

$a(b + c) = \underline{ab + ac}$ **Example:** $4(x + 2) = \underline{4x + 8}$

$(b + c)a = \underline{ba + ca}$ **Example:** $(x + 1)9 = \underline{9x + 9}$

The product of a and $(b - c)$:

$a(b - c) = \underline{ab - ac}$ **Example:** $4(x - 2) = \underline{4x - 8}$

$(b - c)a = \underline{ba - ca}$ **Example:** $(x - 4)6 = \underline{6x - 24}$

Example 2 *Use the Distributive Property with Addition*

Use the distributive property to rewrite the expression without parentheses.

a. $8(x + 9) = \underline{8}(\underline{x}) + \underline{8}(\underline{9})$ **Distribute** $\underline{8}$ **to each term of** $(x + 9)$**.**

$= \underline{8x} + \underline{72}$ **Multiply.**

b. $(7 + y)3 = (\underline{7})\underline{3} + (\underline{y})\underline{3}$ **Distribute** $\underline{3}$ **to each term of** $(7 + y)$**.**

$= \underline{21 + 3y}$ **Multiply.**

Example 3 *Use the Distributive Property with Subtraction*

Use the distributive property to rewrite the expression without parentheses.

a. $5(x - 6) = \underline{5}(\underline{x}) - \underline{5}(\underline{6})$ **Distribute** $\underline{5}$ **to each term of** $(x - 6)$**.**

$= \underline{5x} - \underline{30}$ **Multiply.**

b. $(y - 2)3 = (\underline{y})\underline{3} - (\underline{2})\underline{3}$ **Distribute** $\underline{3}$ **to each term of** $(y - 2)$**.**

$= \underline{3y - 6}$ **Multiply.**

Checkpoint **Use the distributive property to rewrite the expression without parentheses.**

1. $4(a + 9)$ $4a + 36$	**2.** $6(12 + b)$ $72 + 6b$	**3.** $(c + 1)(5)$ $5c + 5$
4. $3(a - 8)$ $3a - 24$	**5.** $9(3 - b)$ $27 - 9b$	**6.** $(c - 12)(3)$ $3c - 36$

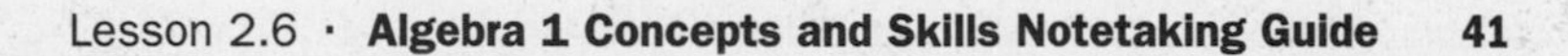

Example 4 *Use the Distributive Property*

A factor with a negative sign must multiply *each* term of an expression. Forgetting to distribute the negative sign to each term is a common error.

Use the distributive property to rewrite the expression without parentheses.

a. $-2(x + 1) = \underline{-2}(x) + (\underline{-2})(1)$ — **Use** distributive property.

$= \underline{-2x - 2}$ — **Multiply.**

b. $(3 + y)(-6) = (3)(\underline{-6}) + (y)(\underline{-6})$ — **Use** distributive property.

$= \underline{-18 - 6y}$ — **Multiply.**

c. $-(x - 1) = \underline{-1}(x) - (\underline{-1})(1)$ — **Use** distributive property.

$= \underline{-x + 1}$ — **Multiply.**

✔ ***Checkpoint*** **Use the distributive property to rewrite the expression without parentheses.**

7. $-3(a + 3)$	8. $(3 + b)(-6)$	9. $-(c - 7)$	10. $(5 - d)(-2)$
$-3a - 9$	$-18 - 6b$	$-c + 7$	$-10 + 2d$

Example 5 *Mental Math Calculations*

Grocery Shopping Cereal is on sale for \$2.53 per box. Use the distributive property and mental math to calculate the cost of 3 boxes of the cereal.

Solution

If you think of \$2.53 as \$2.50 + \$ .03, the mental math is easier.

$3(2.53) = 3(2.50 + \underline{0.03})$ — **Write 2.53 as a sum.**

$= \underline{3}(\underline{2.50}) + \underline{3}(\underline{0.03})$ — **Use distributive property.**

$= \underline{7.50} + \underline{0.09}$ — **Find products mentally.**

$= \underline{7.59}$ — **Find sum mentally.**

Answer The total cost of 3 boxes of cereal is \$ 7.59.

2.7 Combining Like Terms

Goal Simplify an expression by combining terms.

VOCABULARY

Coefficient If a term of an expression consists of a number multiplied by one or more variables, the number is the coefficient of the term. For example, the coefficient of $-3x$ is -3.

Like terms Like terms have the same variables with each variable of the same kind raised to the same power. For example, $8xy$ and $21xy$ are like terms.

Simplified expression An expression is simplified if it has no grouping symbols and if all the like terms have been combined.

Example 1 ***Identify Like Terms***

Note that $-x$ has a coefficient of -1 even though the 1 isn't written. Similarly, x has a coefficient of 1.

Identify the like terms in the expression $x^2 + 3x - 7x^2 - 3 - x$.

Solution

Begin by writing the expression $x^2 + 3x - 7x^2 - 3 - x$ as a sum:

$x^2 + 3x + (\underline{-7x^2}) + (\underline{-3}) + (\underline{-x})$

Answer The terms $\underline{x^2}$ and $\underline{-7x^2}$ are like terms. The terms $\underline{3x}$ and $\underline{-x}$ are also like terms.

Checkpoint **Identify the like terms in the expression.**

1. $5x^2 - 2x + 10x^2$ $5x^2$ and $10x^2$ are like terms.	**2.** $-4b + 9 - 6b$ $-4b$ and $-6b$ are like terms.
3. $13 - 4h^2 - 3h + 8 + 5h$ $-3h$ and $5h$ are like terms. 13 and 8 are like terms.	**4.** $6y + 10 - 5y^2 - 4 - y^2$ $-5y^2$ and $-y^2$ are like terms. 10 and -4 are like terms.

Example 2 ***Combine Like Terms***

Simplify the expression.

a. $7x^2 - 3x^2 = (\underline{7} - \underline{3})x^2$ **Use distributive property.**

$= \underline{4x^2}$ **Add coefficients.**

b. $4x - 2x + 6x - 3$

$= (\underline{4} - \underline{2} + \underline{6})x - 3$ **Use distributive property.**

$= \underline{8x - 3}$ **Add coefficients.**

In Example 2(b), the distributive property has been extended to three terms: $(b + c + d)a = ba + ca + da$.

Example 3 ***Simplify Expressions with Grouping Symbols***

Simplify the expression.

a. $6 - 4(x - 2) = 6 - 4(\underline{x}) - 4(\underline{-2})$ **Use distributive property.**

$= 6 - \underline{4x} + \underline{8}$ **Multiply.**

$= \underline{-4x} + 6 + \underline{8}$ **Group like terms.**

$= \underline{-4x + 14}$ **Combine like terms.**

b. $5(x + 2) - 3(x - 4)$

$= 5(\underline{x}) + 5(\underline{2}) - 3(\underline{x}) - 3(\underline{-4})$ **Use distributive property.**

$= \underline{5x} + \underline{10} - \underline{3x} + \underline{12}$ **Multiply.**

$= \underline{5x} - \underline{3x} + \underline{10} + \underline{12}$ **Group like terms.**

$= \underline{2x + 22}$ **Combine like terms.**

Checkpoint **Simplify the expression.**

5. $4y + 5y - 4$ $9y - 4$	**6.** $6x + 3 - 4x + 2x$ $4x + 3$
7. $7m - 2 + 4(4 - m)$ $3m + 14$	**8.** $-5(n - 3) + 7(1 - n)$ $-12n + 22$

2.8 Dividing Real Numbers

Goal Divide real numbers and use division to simplify algebraic expressions.

VOCABULARY

Reciprocals Two numbers are reciprocals if their product is 1. If $\frac{a}{b}$ is a nonzero number, then its reciprocal is $\frac{b}{a}$.

INVERSE PROPERTY OF MULTIPLICATION

For every nonzero number a, there is a unique number $\frac{1}{a}$ such that: $a \cdot \frac{1}{a} = \underline{1}$ and $\frac{1}{a} \cdot a = \underline{1}$.

Example: $\frac{1}{3} \cdot 3 = \underline{1}$

DIVISION RULE

To divide a number a by a nonzero number b, multiply a by the reciprocal of b. The result is the quotient of a and b.

$$a \div b = a \cdot \frac{1}{b}$$

Example: $-4 \div 7 = -4 \cdot \frac{1}{7} = -\frac{4}{7}$

Example 1 ***Divide Real Numbers***

a. $20 \div (-4) = 20 \cdot \left(-\frac{1}{4}\right) = -\frac{20}{4} = -5$

b. $\frac{5}{2} \div (-25) = \frac{5}{2} \div \left(-\frac{25}{1}\right) = \frac{5}{2} \cdot -\frac{1}{25} = -\frac{1}{10}$

c. $-\frac{3}{4} \div \frac{5}{12} = -\frac{3}{4} \cdot \frac{12}{5} = -\frac{9}{5} = -1\frac{4}{5}$

d. $3\frac{5}{9} \div 2\frac{2}{3} = \frac{32}{9} \div \frac{8}{3} = \frac{32}{9} \cdot \frac{3}{8} = \frac{4}{3} = 1\frac{1}{3}$

When you divide by a mixed number, it is usually easiest to first rewrite the mixed number as an improper fraction.

Example 2 ***Simplify Complex Fractions***

Find the quotient.

A quotient is defined as $\text{quotient} = \frac{\text{dividend}}{\text{divisor}}$. So you can check your solution by showing quotient • divisor = dividend.

a. $\frac{-\frac{2}{5}}{9} = -\frac{2}{5} \div \underline{9} = -\frac{2}{5} \cdot \underline{\frac{1}{9}} = \underline{-\frac{2}{45}}$

b. $\frac{1}{-\frac{5}{8}} = 1 \div \left(\underline{-\frac{5}{8}}\right) = 1 \cdot \left(\underline{-\frac{8}{5}}\right) = \underline{-\frac{8}{5}} = \underline{-1\frac{3}{5}}$

Checkpoint Find the quotient.

1. $-10 \div \frac{1}{3}$ -30	**2.** $3\frac{1}{3} \div 1\frac{1}{4}$ $2\frac{2}{3}$
3. $\frac{\frac{1}{2}}{-5}$ $-\frac{1}{10}$	**4.** $\frac{-6}{\frac{1}{4}}$ -24

THE SIGN OF A QUOTIENT RULE

The quotient of two numbers with the same sign is <u>positive</u>.

$-a \div (-b) = a \div b = \frac{a}{b}$ **Examples:** $-18 \div -3 = \underline{6}$

$18 \div 3 = \underline{6}$

The quotient of two numbers with opposite signs is <u>negative</u>.

$-a \div b = a \div (-b) = -\frac{a}{b}$ **Examples:** $-18 \div 3 = \underline{-6}$

$18 \div -3 = \underline{-6}$

Example 3 *Evaluate an Expression*

Evaluate the expression when $a = 9$ and $b = -4$.

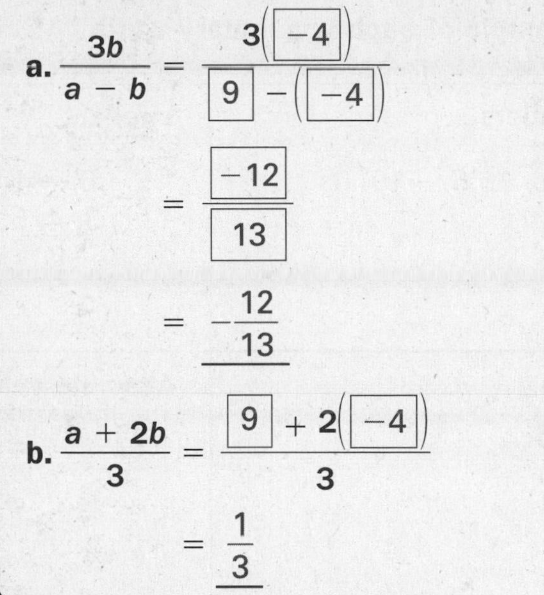

a. $\dfrac{3b}{a-b} = \dfrac{3(\boxed{-4})}{\boxed{9} - (\boxed{-4})}$ — **Substitute** 9 **for *a* and** −4 **for *b*.**

$= \dfrac{\boxed{-12}}{\boxed{13}}$ — **Simplify** numerator **and** denominator**.**

$= -\dfrac{12}{13}$ — **Sign of** quotient **rule**

b. $\dfrac{a+2b}{3} = \dfrac{\boxed{9} + 2(\boxed{-4})}{3}$ — **Substitute** 9 **for *a* and** −4 **for *b*.**

$= \dfrac{1}{3}$ — **Simplify.**

Example 4 *Simplify an Expression*

$\dfrac{56x - 14}{7} = (56x - 14) \div 7$ — **Rewrite fraction as** division **expression.**

$= (56x - 14) \cdot \dfrac{1}{7}$ — **Multiply by** reciprocal**.**

$= (56x)\left(\dfrac{1}{7}\right) - (14)\left(\dfrac{1}{7}\right)$ — **Use** distributive **property.**

$= 8x - 2$ — **Simplify.**

✓ ***Checkpoint*** **Complete the following exercises.**

5. Evaluate $\dfrac{a}{-2b}$ when $a = -\dfrac{1}{5}$ and $b = \dfrac{11}{20}$.	**6.** Simplify $\dfrac{12x - 30}{6}$.
$\dfrac{2}{11}$	$2x - 5$

Words to Review

Give an example of each vocabulary word.

Real numbers $7, \frac{3}{4}, 0.3\overline{8}, 16.5, -10, 0$	**Integers** $\ldots, -5, -4, -3, -2, -1, 0, 1, 2, 3, 4, 5, \ldots$
Opposites -6 and 6	**Absolute value** $\lvert -10 \rvert = 10$
Counterexample The statement "the absolute value of a number is always greater than the number" is false. Counterexample: $\lvert 3 \rvert = 3$.	**Distributive property** $-5(2 + x) = -5(2) + (-5)(x)$
Coefficient 2 is the coefficient of $2x^3$.	**Like terms** $2x^3$ and $-5x^3$
Simplified expression $4x - 5$ is the simplified expression of $\frac{8x - 10}{2}$.	**Reciprocal** The reciprocal of -3 is $-\frac{1}{3}$.

Review your notes and Chapter 2 by using the Chapter Review on pages 121–124 of your textbook.

3.1 Solving Equations Using Addition and Subtraction

Goal Solve linear equations using addition and subtraction.

VOCABULARY

Equivalent equations Equivalent equations are equations that have the same solution(s).

Inverse operations Inverse operations are operations that undo each other, such as addition and subtraction.

Linear equation In a linear equation, the exponent of the variable(s) is one.

TRANSFORMING EQUATIONS

Operation	Original Equation		Equivalent Equation
• Add the same number to *each* side.	$x - 3 = 5$	Add 3.	$x = 8$
• Subtract the same number from *each* side.	$x + 6 = 10$	Subtract 6.	$x = 4$
• Simplify one or both sides.	$x = 8 - 3$	Simplify.	$x = 5$

Example 1 ***Add to Each Side of an Equation***

Solve $x - 9 = -20$

This is a subtraction equation. Use the inverse operation of addition to undo the subtraction.

You can check your solution by substituting your solution for x in the original equation.

$x - 9 = -20$	Write original equation.
$x - 9 + \underline{9} = -20 + \underline{9}$	Add $\underline{9}$ to each side.
$x = \underline{-11}$	Simplify both sides.

Example 2 ***Simplify First***

Solve $n - (-8) = -2$.

$n - (-8) = -2$	Write original equation.
$n + 8 = -2$	Use subtraction rule to simplify.
$n + 8 - \underline{8} = -2 - \underline{8}$	Subtract $\underline{8}$ from each side.
$n = \underline{-10}$	Simplify both sides.

Checkpoint **Solve the equation. Check your solution in the original equation.**

1. $x - 7 = -15$ -8	**2.** $n - (-6) = 4$ -2	**3.** $-7 = 10 + y$ -17
4. $5 - (-z) = 21$ 16	**5.** $m - (-3) = 14$ 11	**6.** $-8 = -b + (-2)$ 6

3.2 Solving Equations Using Multiplication and Division

Goal Solve linear equations using multiplication and division.

VOCABULARY

Properties of equality The rules of algebra used to transform equations into equivalent equations

TRANSFORMING EQUATIONS

Operation	Original Equation		Equivalent Equation
• Multiply *each* side of the equation by the same nonzero number.	$\frac{x}{2} = 3$	Multiply by 2.	$x = 6$
• Divide *each* side of the equation by the same nonzero number.	$4x = 12$	Divide by 4.	$x = 3$

Example 1 ***Divide Each Side of an Equation***

Solve $8x = -3$.

The operation is multiplication. Use the inverse operation of division to isolate the variable x.

$8x = -3$ **Write original equation.**

$\frac{8x}{8} = \frac{-3}{8}$ **Divide each side by 8 to undo the multiplication.**

$x = -\frac{3}{8}$ **Simplify.**

Example 2 *Multiply Each Side of an Equation*

Solve $\frac{x}{-3} = 60$.

The operation is division. Use the inverse operation of multiplication to isolate the variable x.

When you multiply or divide both sides of an equation by a negative number, be careful with the signs of the numbers.

$\frac{x}{-3} = 60$	**Write original equation.**
$\underline{(-3)}\left(\frac{x}{-3}\right) = \underline{(-3)}(60)$	**Multiply each side by $\underline{-3}$ to undo the division.**
$x = \underline{-180}$	**Simplify.**

Answer The solution is $\underline{-180}$.

Checkpoint **Solve the equation. Check your solution in the original equation.**

1. $\frac{n}{-4} = 7$ -28	**2.** $6 = \frac{x}{7}$ 42
3. $63 = 9x$ 7	**4.** $-9r = -54$ 6

Example 3 *Multiply Each Side by a Reciprocal*

When you solve an equation with a fractional coefficient, such as $-\frac{3}{4}m = 15$, you can isolate the variable by multiplying by the reciprocal of the fraction.

Solve $-\frac{3}{4}m = 15$.

The fractional coefficient is $-\frac{3}{4}$. The reciprocal of $-\frac{3}{4}$ is $-\frac{4}{3}$.

$-\frac{3}{4}m = 15$ **Write original equation.**

$\left(-\frac{4}{3}\right)\left(-\frac{3}{4}m\right) = \left(-\frac{4}{3}\right)15$ **Multiply each side by the reciprocal, $-\frac{4}{3}$.**

$m = -20$ **Simplify.**

Answer The solution is -20.

PROPERTIES OF EQUALITY

Addition Property of Equality	If $a = b$, then $a + c = b + c$.
Subtraction Property of Equality	If $a = b$, then $a - c = b - c$.
Multiplication Property of Equality	If $a = b$, then $ca = cb$.
Division Property of Equality	If $a = b$ and $c \neq 0$, then $\frac{a}{c} = \frac{b}{c}$.

Checkpoint **Solve the equation. Check your solution in the original equation.**

5. $\frac{4}{9}x = 12$ 27	**6.** $15 = -\frac{5}{3}x$ -9

3.3 Solving Multi-Step Equations

Goal Use two or more steps to solve a linear equation.

Example 1 ***Solve a Linear Equation***

Solve $2x - 4 = -18$.

Solution

To isolate the variable, undo the subtraction and then the multiplication.

$2x - 4 = -18$	**Write original equation.**
$2x - 4 + \underline{4} = -18 + \underline{4}$	**Add 4 to each side to undo the subtraction.**
$2x = \underline{-14}$	**Simplify both sides.**
$\frac{2x}{2} = \frac{-14}{2}$	**Divide each side by 2 to undo the multiplication.**
$x = \underline{-7}$	**Simplify.**

Example 2 ***Combine Like Terms First***

Solve $8x - 5x + 16 = -29$.

Solution

$8x - 5x + 16 = -29$	**Write original equation.**
$\underline{3x} + 16 = -29$	**Combine like terms $8x$ and $-5x$.**
$\underline{3x} + 16 - \underline{16} = -29 - \underline{16}$	**Subtract 16 from each side to undo the addition.**
$\underline{3x} = \underline{-45}$	**Simplify both sides.**
$\frac{3x}{3} = \frac{-45}{3}$	**Divide each side by 3 to undo the multiplication.**
$x = \underline{-15}$	**Simplify.**

Checkpoint Solve the equation. Check your solution in the original equation.

1. $2x - 5 = 9$ 7	2. $3 - 4a = 19$ -4
3. $12m - 4m + 3 = -29$ -4	4. $35 = 7y + 13y - 5$ 2

Example 3 ***Use the Distributive Property***

Solve $9x - 5(x + 6) = -14$.

Remember to distribute the negative sign to *each* term inside the parentheses, not just the first term.

$9x - 5(x + 6) = -14$	**Write original equation.**
$9x - 5x - 30 = -14$	**Use distributive property.**
$4x - 30 = -14$	**Combine like terms $9x$ and $-5x$.**
$4x - 30 + 30 = -14 + 30$	**Add 30 to each side to undo the subtraction.**
$4x = 16$	**Simplify.**
$\frac{4x}{4} = \frac{16}{4}$	**Divide each side by 4 to undo the multiplication.**
$x = 4$	**Simplify.**

Example 4 *Multiply by a Reciprocal First*

Solve $24 = \frac{3}{4}(x + 7)$.

$24 = \frac{3}{4}(x + 7)$	**Write original equation.**
$\frac{4}{3}(24) = \frac{4}{3}\left(\frac{3}{4}\right)(x + 7)$	**Multiply each side by $\frac{4}{3}$, the reciprocal of $\frac{3}{4}$.**
$32 = x + 7$	**Simplify.**
$32 - 7 = x + 7 - 7$	**Subtract 7 from each side.**
$25 = x$	**Simplify both sides.**

Checkpoint **Solve the equation. Check your solution in the original equation.**

5. $-2(3 - k) = 30$ 18	**6.** $-38 = 4(n - 2) + 2n$ -5
7. $\frac{2}{5}(j + 23) = 8$ -3	**8.** $12 = \frac{1}{3}(g + 2)$ 34

3.4 Solving Equations with Variables on Both Sides

Goal Solve equations that have variables on both sides.

VOCABULARY

Identity An identity is an equation that is true for all values of the variable.

Example 1 ***Collect Variables on One Side***

Solve $4x - 10 = 32 - 3x$.

Solution

Look at the coefficients of the x-terms. Because 4 is greater than -3, collect the x-terms on the left side to get a positive coefficient.

$4x - 10 = 32 - 3x$	**Write original equation.**
$4x - 10 + \underline{3x} = 32 - 3x + \underline{3x}$	**Add $\underline{3x}$ to each side.**
$\underline{7x} - 10 = 32$	**Combine like terms.**
$\underline{7x} - 10 + \underline{10} = 32 + \underline{10}$	**Add $\underline{10}$ to each side.**
$\underline{7x} = \underline{42}$	**Simplify both sides.**
$\frac{7x}{\underline{7}} = \frac{42}{\underline{7}}$	**Divide each side by $\underline{7}$.**
$x = \underline{6}$	**Simplify.**

Answer The solution is $\underline{6}$.

Check $4x - 10 = 32 - 3x$	**Write original equation.**
$4(\underline{6}) - 10 \stackrel{?}{=} 32 - 3(\underline{6})$	**Substitute $\underline{6}$ for each x.**
$\underline{14} = \underline{14}$	**Solution is $\underline{\text{correct}}$.**

Example 2 *Combine Like Terms First*

Solve $2x - 9 + 7x = 4x - 34$.

$2x - 9 + 7x = 4x - 34$	**Write original equation.**
$\underline{9x} - 9 = 4x - 34$	**Combine like terms.**
$\underline{9x} - 9 - \underline{4x} = 4x - 34 - \underline{4x}$	**Subtract $\underline{4x}$ from each side.**
$\underline{5x} - 9 = -34$	**Combine like terms.**
$\underline{5x} - 9 + \underline{9} = -34 + \underline{9}$	**Add $\underline{9}$ to each side.**
$\underline{5x} = \underline{-25}$	**Simplify both sides.**
$\dfrac{5x}{5} = \dfrac{-25}{5}$	**Divide each side by $\underline{5}$.**
$x = \underline{-5}$	**Simplify.**

Answer The solution is $\underline{-5}$. Check this in the original equation.

Checkpoint **Solve the equation. Check your solution in the original equation.**

1. $6x = 5x - 33$ -33	**2.** $10p - 22 = -p$ 2
3. $8k - 22 = 10k$ -11	**4.** $2n - 4n = 3n + 17$ $-3\frac{2}{5}$

Example 3 *Identify Number of Solutions*

Solve the equation if possible. Determine whether it has *one solution*, *no solution*, or is an *identity*.

a. $2(4x + 5) = 8x + 10$ **b.** $x - 1 = x + 7$

Solution

a.

$2(4x + 5) = 8x + 10$	**Write original equation.**
$8x + 10 = 8x + 10$	**Use distributive property.**
$8x + 10 - 8x = 8x + 10 - 8x$	**Subtract $8x$ from each side.**
$10 = 10$	**Combine like terms.**

Answer The equation $10 = 10$ is always true, so all values of x are solutions. The original equation is an identity.

b.

$x - 1 = x + 7$	**Write original equation.**
$x - 1 - x = x + 7 - x$	**Subtract x from each side.**
$-1 \neq 7$	**Combine like terms.**

Answer The equation $-1 = 7$ is never true no matter what the value of x. The original equation has no solution.

✔ ***Checkpoint*** **Solve the equation if possible. Determine whether the equation has *one solution*, *no solution*, or is an *identity*.**

5. $3(x - 2) = 3x - 6$	**6.** $3(x - 2) = 3x + 1$
identity	no solution

3.5 More on Linear Equations

Goal Solve more complicated equations that have variables on both sides.

STEPS FOR SOLVING LINEAR EQUATIONS

1. **Simplify** each side by distributing and/or combining like terms.
2. **Collect** variable terms on the side where the coefficient is greater.
3. **Use** inverse operations to isolate the variable.
4. **Check** your solution in the original equation.

Example 1 ***Solve a More Complicated Equation***

Solve $3(2 - x) - x = -5(x + 1)$.

Solution

You should simplify an equation before deciding whether to collect the variable terms on the right side or the left side.

$3(2 - x) - x = -5(x + 1)$	**Write original equation.**
$6 - 3x - x = -5x - 5$	**Use distributive property.**
$6 - 4x = -5x - 5$	**Combine like terms.**
$6 - 4x + 5x = -5x - 5 + 5x$	**Add** $5x$ **to each side.**
$6 + x = -5$	**Combine like terms.**
$6 + x - 6 = -5 - 6$	**Subtract** 6 **from each side.**
$x = -11$	**Simplify.**

Answer The solution is -11.

Check $3(2 - x) - x = -5(x + 1)$	**Write original equation.**
$3(2 - (-11)) - (-11) \stackrel{?}{=} -5(-11 + 1)$	**Substitute** -11 **for each** x.
$3(13) + 11 \stackrel{?}{=} -5(-10)$	**Simplify.**
$39 + 11 \stackrel{?}{=} 50$	**Multiply.**
$50 = 50$	**Solution is** correct.

Example 2 *Solve a More Complicated Equation*

Solve $2(5 - 4x) = 9(x + 10) - 7x$.

$2(5 - 4x) = 9(x + 10) - 7x$	**Write original equation.**
$10 - 8x = 9x + 90 - 7x$	**Use distributive property.**
$10 - 8x = 2x + 90$	**Combine like terms.**
$10 = 10x + 90$	**Add** $8x$ **to each side.**
$-80 = 10x$	**Subtract** 90 **from each side.**
$x = -8$	**Divide each side by** 10.

You can use mental math to add $8x$ to each side of the equation.

Example 3 *Solve a More Complicated Equation*

Solve $\frac{1}{5}(15x + 20) = 6 - 2(x - 4)$.

$\frac{1}{5}(15x + 20) = 6 - 2(x - 4)$	**Write original equation.**
$3x + 4 = 6 - 2x + 8$	**Use distributive property.**
$3x + 4 = 14 - 2x$	**Simplify.**
$5x + 4 = 14$	**Add** $2x$ **to each side.**
$5x = 10$	**Subtract** 4 **from each side.**
$x = 2$	**Divide each side by** 5.

Checkpoint **Solve the equation. Check your solution in the original equation.**

1. $-6(4 - x) = 12x - 24$ 0	**2.** $-\frac{1}{6}(30 - 12p) = 4 - 3(p - 2)$ 3

Example 4 *Compare Payment Plans*

Golf Course Fees A private golf course charges $1200 for membership and $5 per round played to be a member. A guest of a member pays $45 per round. Compare the costs of the members and guests.

Solution

Find the number of rounds for which the costs would be the same.

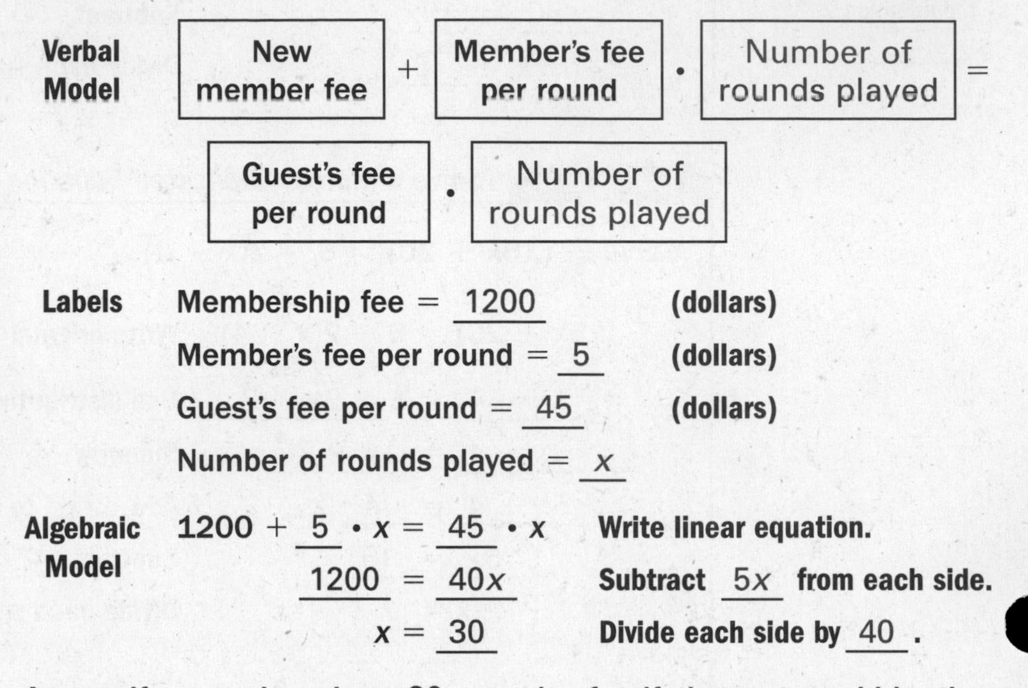

Algebraic Model

$1200 + 5 \cdot x = 45 \cdot x$ **Write linear equation.**

$1200 = 40x$ **Subtract $5x$ from each side.**

$x = 30$ **Divide each side by 40.**

Answer If a member plays 30 rounds of golf, the cost would be the same as a guest playing 30 rounds. If a member plays more than 30 rounds, it would cost less to be a member. If a member plays less than 30 rounds, it would cost less to be a guest.

Checkpoint **Complete the following exercise.**

3. You are looking to rent a banquet hall for a birthday party. Banquet hall A charges $200 for set-up and rental plus $15 per person. Banquet hall B charges $275 for set-up and rental plus $12 per person. Compare the costs of banquet halls A and B.

If you have 25 guests, the cost would be the same for both halls. If you have more than 25 guests, banquet hall B would be cheaper.

Solving Decimal Equations

Goal Find exact and approximate solutions of equations that contain decimals.

VOCABULARY

Rounding error Rounding error is the error produced when a decimal expansion is limited to a specific number of digits to the right of the decimal point.

Example 1 ***Round for the Final Answer***

Solve $-28x + 31 = 124$. Round to the nearest hundredth.

Solution

$-28x + 31 = 124$	**Write original equation.**
$-28x = \underline{93}$	**Subtract** $\underline{31}$ **from each side.**
$x = \dfrac{93}{-28}$	**Divide each side by** $\underline{-28}$.
$x \approx \underline{-3.321428571}$	**Use a calculator to get an approximate solution.**
$x \approx \underline{-3.32}$	**Round to nearest hundredth.**

Answer The solution is approximately $\underline{-3.32}$.

Check

$-28x + 31 = 124$	**Write original equation.**
$-28(\underline{-3.32}) + 31 \stackrel{?}{=} 124$	**Substitute** $\underline{-3.32}$ **for** ***x***.
$\underline{123.96} \approx 124$	**Rounded answer is** $\underline{\text{reasonable}}$.

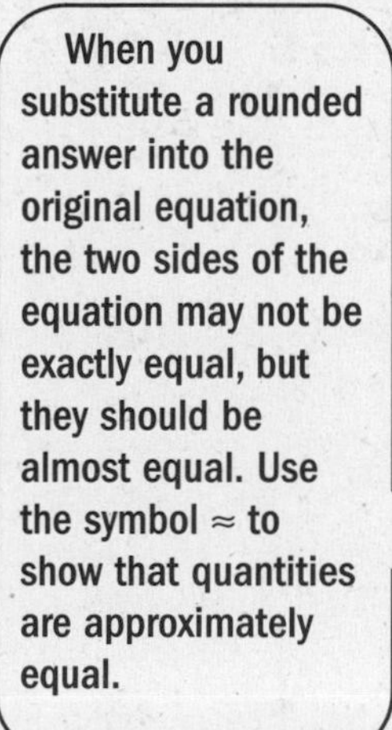

Example 2 *Solve an Equation that Contains Decimals*

Solve $2.5x - 60.2 = 0.2x$. Round to the nearest tenth.

$2.5x - 60.2 = 0.2x$	**Write original equation.**
$\underline{2.3x} - 60.2 = 0$	**Subtract $\underline{0.2x}$ from each side.**
$\underline{2.3x} = \underline{60.2}$	**Add $\underline{60.2}$ to each side.**
$x = \underline{\frac{60.2}{2.3}}$	**Divide each side by $\underline{2.3}$.**
$x \approx \underline{26.17391304}$	**Use a calculator to get an approximate solution.**
$x \approx \underline{26.2}$	**Round to nearest tenth.**

Answer The solution is approximately $\underline{26.2}$.

✔ ***Checkpoint*** **Solve the equation. Round to the nearest hundredth.**

1. $11x - 5 = 26$	**2.** $23 - 6y = 7$	**3.** $-48 = 13n + 14$
2.82	2.67	-4.77

Solve the equation. Round to the nearest tenth.

4. $6.3x - 54.8 = 0.8x + 9.5$	**5.** $12.8 + 2.7x = 5.5 + 7.2x$
11.7	1.6

Example 3 *Round for a Practical Answer*

Three people are equally sharing the cost of a monthly cable bill. The cable bill for this month is $41.35. What is each person's share?

Solution

You can find each person's share for this month's cable bill by solving $3x = 41.35$.

$3x = 41.35$	**Write original equation.**
$x = \underline{13.78333\ldots}$	**Use a calculator to divide each side by _3_. Exact answer is a repeating decimal.**
$x \approx \underline{13.78}$	**Round to nearest cent.**

Answer Each person's share of the cable bill is $ _13.78_.

Three times the rounded answer is _one cent_ too _little_ due to _rounding error_.

✓ *Checkpoint* **Complete the following exercise.**

6. Six friends are equally sharing the cost of two pizzas. The total cost of the pizzas is $18.95. What does each person owe?

$3.16

3.7 Formulas

Goal Solve a formula for one of its variables.

VOCABULARY

Formula A formula is an algebraic equation that relates two or more quantities.

Example 1 ***Solve a Temperature Conversion Formula***

In the formula $F = \frac{9}{5}C + 32$, F represents degrees Fahrenheit and C represents degrees Celsius. Solve the formula for C.

Solution

To solve for the variable C, transform the original formula to isolate C. Use the steps for solving a linear equation.

$F = \frac{9}{5}C + 32$ — **Write original formula.**

$F - \underline{32} = \frac{9}{5}C + 32 - \underline{32}$ — **Subtract $\underline{32}$ from each side.**

$F - \underline{32} = \frac{9}{5}C$ — **Simplify.**

$\underline{\frac{5}{9}} \cdot (F - \underline{32}) = \underline{\frac{5}{9}} \cdot \frac{9}{5}C$ — **Multiply each side by $\underline{\frac{5}{9}}$, the reciprocal of $\frac{9}{5}$.**

$\underline{\frac{5}{9}}(F - \underline{32}) = C$ — **Simplify.**

Answer The new formula is $C = \underline{\frac{5}{9}}(F - \underline{32})$.

Example 2 *Solve and Use an Area Formula*

The formula for the area of a rectangle is $A = \ell w$.

a. Find a formula for width w in terms of area A and length ℓ.

b. Use the new formula to find the width of a rectangle that has an area of 54 square inches and a length of 9 inches.

Solution

a. Solve for width w.

$A = \ell w$ **Write original formula.**

$\frac{A}{\ell} = w$ **Divide each side by ℓ.**

b. Substitute the given values into the new formula.

$w = \frac{A}{\ell} = \frac{54}{9} = 6$ inches

Example 3 *Solve and Use a Density Formula*

The density d of a substance is found by dividing its mass m by its volume v.

a. Solve the density formula $d = \frac{m}{v}$ for volume v.

b. Use the new formula to find the volume of a substance that has a density of 7.2 grams per cubic centimeter and a mass of 25.2 grams.

Solution

a. Solve for volume v.

$d = \frac{m}{v}$ **Write original formula.**

$dv = m$ **Multiply each side by v.**

$v = \frac{m}{d}$ **Divide each side by d.**

b. Substitute the given values into the new formula.

$v = \frac{m}{d} = \frac{25.2}{7.2} = 3.5$ cubic centimeters

Example 4 *Solve and Use a Distance Formula*

The distance traveled d is found by multiplying the rate (or speed) r by the time t.

a. Solve the distance formula $d = rt$ for time t.

b. A car travels 1770 miles on a highway at an average speed of 59 miles per hour. How long does the trip take?

Solution

a. $d = rt$ **Write original formula.**

$\frac{d}{r} = \frac{rt}{r}$ **Divide each side by** r.

$\frac{d}{r} = t$ **Simplify.**

b. $t = \frac{d}{r} = \frac{1770}{59} = 30$

Answer The trip takes 30 hours.

✓ *Checkpoint* **Complete the following exercises.**

1. The perimeter P of a rectangle with length ℓ and width w is $P = 2\ell + 2w$. Solve the equation for ℓ.

   $\ell = \frac{P - 2w}{2}$

2. The volume V of a box with length ℓ, width w, and height h is $V = \ell wh$. Solve the equation for w.

   $w = \frac{V}{\ell h}$

3. Use the result from Exercise 2 to find the width w of a box with a length of 2 feet, a height of 3 feet, and a volume of 12 cubic feet.

   2 feet

3.8 Ratios and Rates

Goal Use ratios and rates to solve real-life problems.

VOCABULARY

Ratio of *a* to *b* The ratio of a to b is the relationship $\frac{a}{b}$ of two quantities a and b.

Rate of *a* per *b* The rate of a per b is the relationship $\frac{a}{b}$ of two quantities a and b that are measured in different units.

Unit rate A unit rate is a rate per one given unit, such as 60 miles per gallon.

Unit analysis Unit analysis is writing the units of each variable in a real-life problem to help determine the units for the answer.

Example 1 ***Find a Ratio***

A baseball player has 16 hits in 44 at bats. Find the ratio of hits to at bats.

$$\text{Ratio} = \frac{\text{hits}}{\text{at bats}} = \frac{16}{44} = \frac{4}{11}$$

Answer **The ratio is $\frac{4}{11}$, which is read as "4 to 11."**

> A ratio compares two quantities measured in the *same* unit. The ratio itself has no units. A rate compares two quantities that have *different* units.

Example 2 ***Find a Unit Rate***

A car travels 330 miles in 6 hours. What is the average speed of the car in miles per hour?

$$\text{Rate} = \frac{330\ \boxed{\text{mi}}}{6\ \boxed{\text{h}}} = \frac{55\text{ mi}}{1\text{ h}} = 55\text{ mi/h}$$

Answer **The average speed of the car is 55 miles per hour.**

 ***Checkpoint*** **Complete the following exercises.**

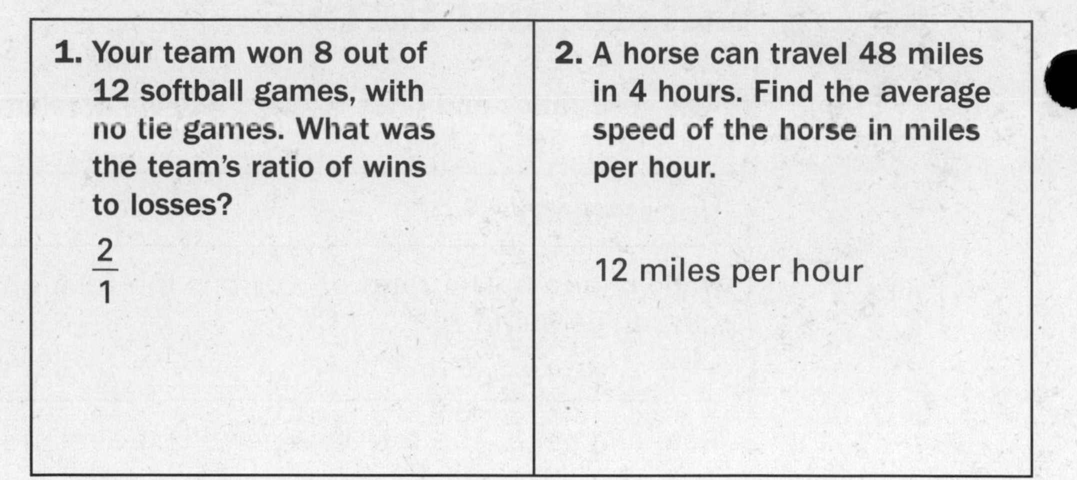

1. Your team won 8 out of 12 softball games, with no tie games. What was the team's ratio of wins to losses? $\frac{2}{1}$	**2.** A horse can travel 48 miles in 4 hours. Find the average speed of the horse in miles per hour. 12 miles per hour

Example 3 ***Use Unit Analysis***

Use unit analysis to convert 6 minutes to seconds.

Solution

Use the fact that 60 seconds = 1 minute.

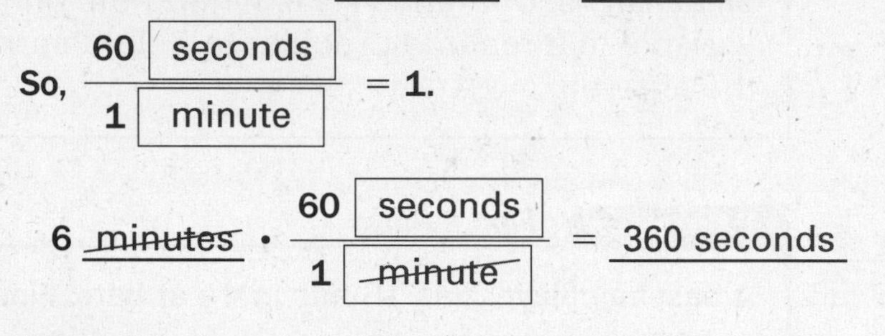

So, $\frac{60 \text{ seconds}}{1 \text{ minute}} = 1.$

$6 \text{ }\cancel{\text{minutes}} \cdot \frac{60 \text{ seconds}}{1 \text{ }\cancel{\text{minute}}} = 360 \text{ seconds}$

Answer 6 minutes equals 360 seconds.

Example 4 ***Use a Rate***

The average mileage for your car is 34 miles per gallon of gasoline. How many miles can you drive on a full 14 gallon tank of gasoline?

Solution

$\text{distance} = \left(34 \frac{\text{mi}}{\cancel{\text{gal}}}\right)(14 \text{ }\cancel{\text{gal}})$	**Substitute rate and gallons.**
$= (34 \text{ mi})(14)$	**Use unit analysis.**
$= 476 \text{ mi}$	**Multiply.**

Answer You can drive about 476 miles on 14 gallons.

Example 5 *Apply Unit Analysis*

Exchanging Money You are visiting Finland and you want to exchange \$350 for euros. The rate of currency exchange is 0.92 euro per United States dollar when you exchange the money. How many euros will you receive?

Solution

You can use unit analysis to write an equation to convert dollars into euros.

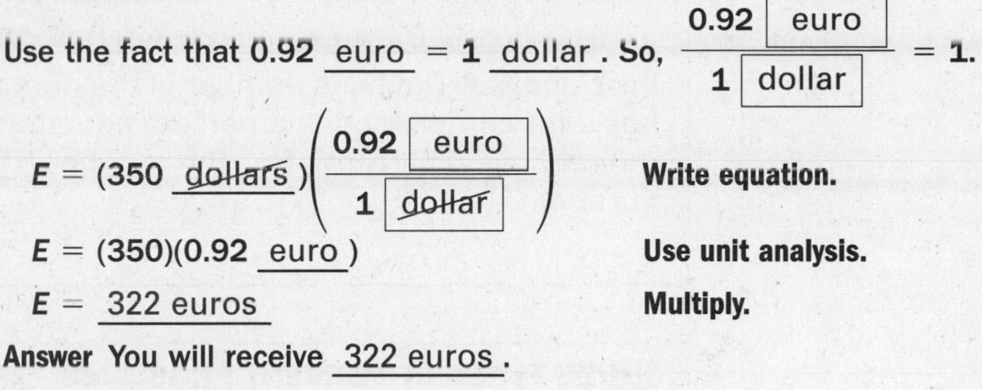

Use the fact that 0.92 euro = 1 dollar. So, $\frac{0.92 \text{ euro}}{1 \text{ dollar}} = 1$.

$E = (350 \cancel{\text{dollars}})\left(\frac{0.92 \text{ euro}}{1 \cancel{\text{dollar}}}\right)$ **Write equation.**

$E = (350)(0.92 \text{ euro})$ **Use unit analysis.**

$E = 322 \text{ euros}$ **Multiply.**

Answer You will receive 322 euros.

Checkpoint Complete the following exercises.

3. Use unit analysis to convert 3 years to days.

1095 days

4. A lawnmower can mow about 1.25 acres per gallon of fuel. How many acres are mowed after using 6 gallons of fuel?

7.5 acres

5. In Example 5, you have 34 euros when you leave Finland. The rate of currency exchange is now 0.93 euros per United States dollar. How many United States dollars will you get back? Round to the nearest cent.

\$36.56

3.9 Percents

Goal Solve percent problems.

VOCABULARY

Percent A percent is a ratio that compares a number to 100.

Base number The base number is the number that is the basis for comparison in a percent equation. In the verbal model "a is p percent of b" the number b is the base number.

THREE TYPES OF PERCENT PROBLEMS

Question	Given	Need to Find
What is p percent of b?	b and p	Number compared to base, a
a is p percent of what?	a and p	Base number, b
a is what percent of b?	a and b	Percent, p

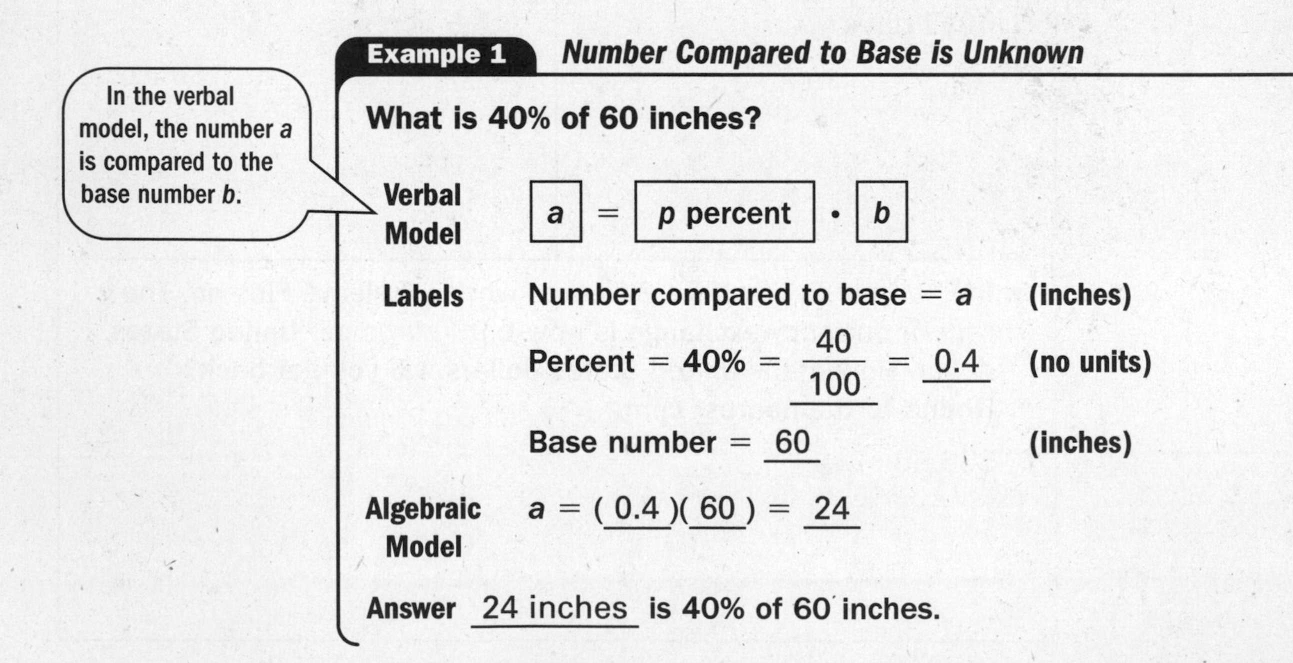

In the verbal model, the number a is compared to the base number b.

Example 1 ***Number Compared to Base is Unknown***

What is 40% of 60 inches?

Verbal Model $\boxed{a} = \boxed{p \text{ percent}} \cdot \boxed{b}$

Labels Number compared to base $= a$ **(inches)**

Percent $= 40\% = \frac{40}{100} = 0.4$ **(no units)**

Base number $= 60$ **(inches)**

Algebraic Model $a = (0.4)(60) = 24$

Answer 24 inches is 40% of 60 inches.

Example 2 ***Base Number is Unknown***

Twelve dollars is 15% of what amount of money?

Verbal Model $\boxed{a} = \boxed{p \text{ percent}} \cdot \boxed{b}$

Labels Number compared to base = 12 **(dollars)**

Percent = 15% = $\frac{15}{100}$ = $\frac{3}{20}$ **(no units)**

Base number = b **(dollars)**

Algebraic Model $12 = \left(\frac{3}{20}\right)b$

$80 = b$

Answer $12 is 15% of $80.

Example 3 ***Percent is Unknown***

Two hundred fifty-five is what percent of 85?

Verbal Model $\boxed{a} = \boxed{p \text{ percent}} \cdot \boxed{b}$

Labels Number compared to base = 255 **(no units)**

Percent = $p\%$ = $\frac{p}{100}$ **(no units)**

Base number = 85 **(no units)**

Algebraic Model $255 = \left(\frac{p}{100}\right)(85)$

$\frac{255}{85} = \frac{p}{100}$

$3 = \frac{p}{100}$

$300 = p$

Answer 255 is 300% of 85.

 ***Checkpoint*** **Complete the following exercises.**

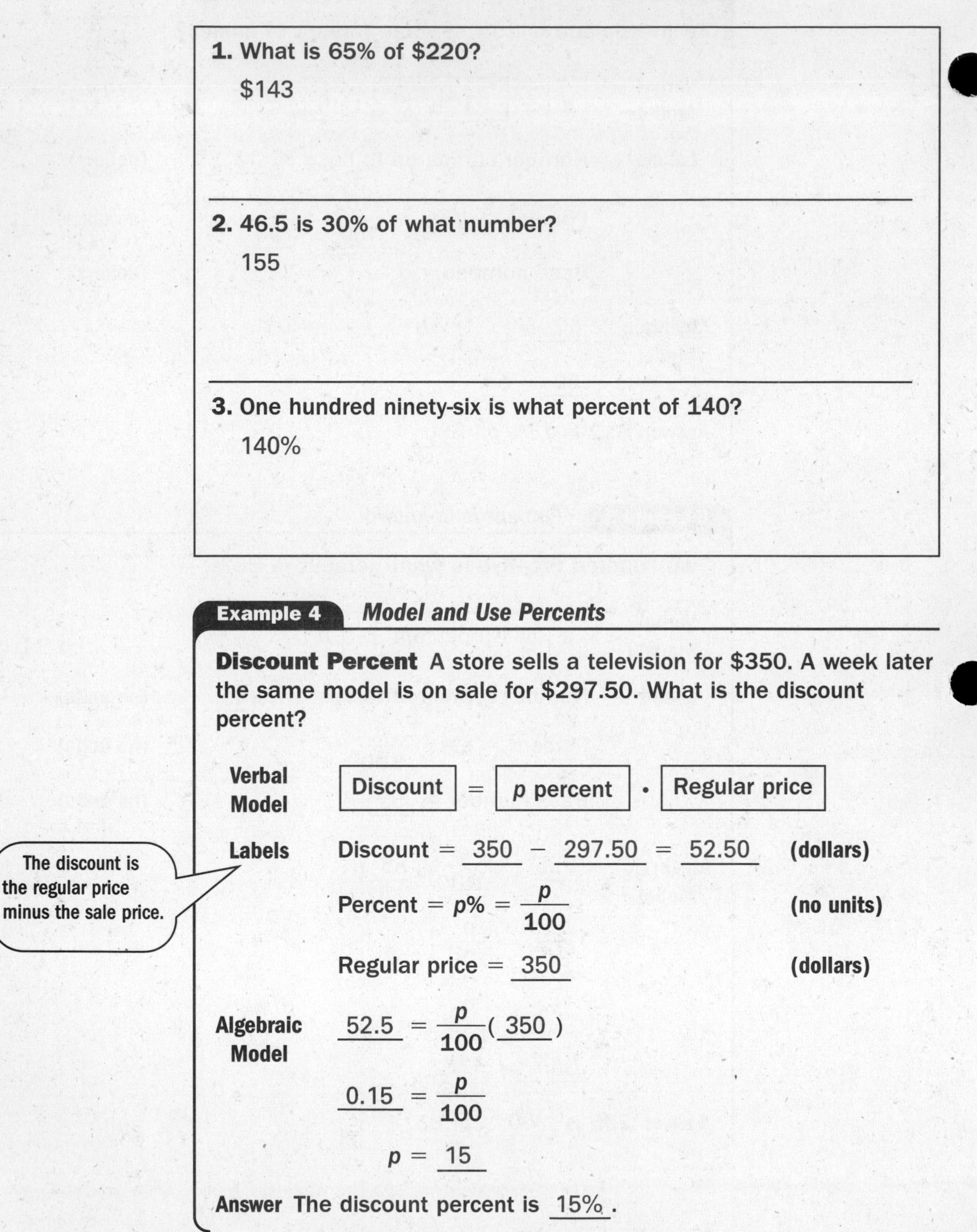

1. What is 65% of $220?

$143

2. 46.5 is 30% of what number?

155

3. One hundred ninety-six is what percent of 140?

140%

Example 4 ***Model and Use Percents***

Discount Percent A store sells a television for $350. A week later the same model is on sale for $297.50. What is the discount percent?

Verbal Model | Discount | = | *p* percent | • | Regular price |

Labels Discount = 350 − 297.50 = 52.50 **(dollars)**

Percent $= p\% = \frac{p}{100}$ **(no units)**

Regular price = 350 **(dollars)**

> The discount is the regular price minus the sale price.

Algebraic Model

$52.5 = \frac{p}{100}(350)$

$0.15 = \frac{p}{100}$

$p = 15$

Answer The discount percent is 15%.

Words to Review

Give an example of the vocabulary word.

Equivalent equations $x - 9 = 3$ and $x = 12$ are equivalent equations.	**Inverse operations** Addition and subtraction
Linear equation $y = 5x - 67$	**Properties of equality** The Multiplication Property of Equality is as follows: If $a = b$, then $ca = cb$.
Identity $2(x - 1) = 2x - 2$	**Rounding error** When dividing a lunch bill of \$18.67 among 3 people, the share rounds to \$6.22. Three times this rounded answer is one cent too little.
Formula The distance formula, $d = rt$	**Ratio of *a* to *b*** Wins to losses, $\frac{\text{6 games won}}{\text{8 games lost}}$
Rate of *a* per *b* $\frac{\text{122 miles}}{\text{4 hours}}$	**Unit rate** $\frac{\text{56 miles}}{\text{1 hour}}$

Unit analysis	Percent
$\frac{\text{dollars}}{\cancel{\text{gallon}}} \cdot \cancel{\text{gallon}} = \text{dollars}$	12%

Base number
24 is the base number in the following: 8 is what percent of 24?

Review your notes and Chapter 3 by using the Chapter Review on pages 189–192 of your textbook.

4.1 The Coordinate Plane

Goal Plot points in a coordinate plane.

VOCABULARY

Coordinate plane The coordinate plane is the coordinate system formed by two real number lines that intersect at a right angle.

Origin The origin is the point in a coordinate plane where the horizontal axis intersects the vertical axis.

***x*-axis** The x-axis is the horizontal axis in a coordinate plane.

***y*-axis** The y-axis is the vertical axis in a coordinate plane.

Ordered pair An ordered pair is a pair of numbers used to identify a point in a coordinate plane.

***x*-coordinate** The x-coordinate is the first number in an ordered pair.

***y*-coordinate** The y-coordinate is the second number in an ordered pair.

Quadrant A quadrant is one of four regions into which the axes divide a coordinate plane.

Scatter plot A scatter plot is a coordinate graph containing points that represent a set of ordered points. A scatter plot is used to analyze relationships between two real-life quantities.

Example 1 *Plot Points in a Coordinate Plane*

Plot the points $A(-2, 3)$, $B(3, -4)$, and $C(0, -2)$ in a coordinate plane.

To plot the point $A(-2, 3)$, start at the origin. Move 2 units to the left and 3 units up.

To plot the point $B(3, -4)$, start at the origin. Move 3 units to the right and 4 units down.

To plot the point $C(0, -2)$, start at the origin. Move 0 units to the left and 2 units down.

A(−2, 3)
C(0, −2)
B(3, −4)

Example 2 *Identify Quadrants*

Name the quadrants the points $D(-2, -9)$ and $E(12, 4)$ are in.

The point $D(-2, -9)$ is in Quadrant III because its x- and y-coordinates are both negative.

The point $E(12, 4)$ is in Quadrant I because its x- and y-coordinates are both positive.

Checkpoint **Plot the points in the same coordinate plane.**

1. $A(-3, -2)$ **2.** $B(4, 0)$

3. $C(1, 4)$ **4.** $D(-3, 2)$

C(1, 4)
D(−3, 2)
B(4, 0)
A(−3, −2)

Name the quadrant the point is in.

5. $(-6, 7)$	**6.** $(-6, -7)$	**7.** $(6, -7)$	**8.** $(6, 7)$
Quadrant II	Quadrant III	Quadrant IV	Quadrant I

Example 3 *Make a Scatter Plot*

NCAA Basketball Teams The number of NCAA men's college basketball teams is shown in the table.

Year	1995	1996	1997	1998	1999	2000
Men's teams	868	866	865	895	926	932

a. Make a scatter plot of the data.

b. Describe the pattern of the number of men's basketball teams.

Solution

a. Let M represent the number of men's teams. Let t represent the number of years since 1995.

Because you want to see how the number of teams changed over time, put t on the horizontal axis and M on the vertical axis.

Choose a scale. Use a break in the scale for the number of teams to focus on the values between 800 and 950.

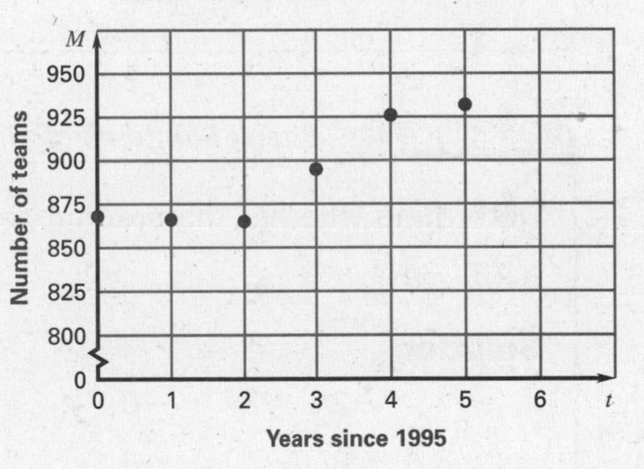

b. From the scatter plot, you can see that the number of men's teams in the NCAA was fairly constant for three years and then began to increase gradually.

4.2 Graphing Linear Equations

Goal Graph a linear equation using a table of values.

VOCABULARY

Linear equation A linear equation in two variables x and y is an equation that can be written in the form $Ax + By = C$, where A and B are not both zero.

Solution of an equation A solution of an equation in two variables x and y is an ordered pair (x, y) that makes the equation true.

Function form A two-variable equation is in function form if one of its variables is isolated on one side of the equation.

Graph of an equation The graph of an equation in two variables x and y is the set of all points (x, y) that are solutions of the equation.

Example 1 ***Check Solutions of Linear Equations***

Determine whether the ordered pair is a solution of $2x + 3y = -6$.

a. $(3, -4)$ **b.** $(-4, 1)$

Solution

a. $2x + 3y = -6$ **Write original equation.**

$2(\underline{3}) + 3(\underline{-4}) \stackrel{?}{=} -6$ **Substitute** $\underline{3}$ **for x and** $\underline{-4}$ **for y.**

$\underline{-6}\ \underline{=}\ -6$ **Simplify.** $\underline{\text{True}}$ **statement.**

Answer $(3, -4)$ $\underline{\text{is}}$ a solution of the equation $2x + 3y = -6$.

b. $2x + 3y = -6$ **Write original equation.**

$2(\underline{-4}) + 3(\underline{1}) \stackrel{?}{=} -6$ **Substitute** $\underline{-4}$ **for x and** $\underline{1}$ **for y.**

$\underline{-5}\ \underline{\neq}\ -6$ **Simplify.** $\underline{\text{Not a true}}$ **statement.**

Answer $(-4, 1)$ $\underline{\text{is not}}$ a solution of the equation $2x + 3y = -6$.

Checkpoint **Determine whether the ordered pair is a solution of $-2x + y = 3$.**

1. (0, 3)	**2.** (1, 1)	**3.** (1, 5)
Solution	Not a solution	Solution

Example 2 ***Find Solutions of Linear Equations***

Find three ordered pairs that are solutions of $-5x + y = -2$.

1. Rewrite the equation in function form to make it easier to substitute values into the equation.

$-5x + y = -2$ **Write original equation.**

$y = \underline{5x - 2}$ **Add $\underline{5x}$ to each side.**

2. Choose any value for x and substitute it into the equation to find the corresponding y-value. The easiest x-value is $\underline{0}$.

$y = 5(\underline{0}) - 2$ **Substitute $\underline{0}$ for x.**

$y = \underline{-2}$ **Simplify. The solution is $\underline{(0, -2)}$.**

3. Select a few more values of x and make a table to record the solutions.

x	0	1	2	3	−1	−2
y	−2	3	8	13	−7	−12

Answer $\underline{(0, -2)}$, $\underline{(1, 3)}$, and $\underline{(-1, -7)}$ are three solutions of $-5x + y = -2$.

GRAPHING A LINEAR EQUATION

Step 1 Rewrite the equation in function form, if necessary.

Step 2 Choose a few values of x and make a table of values.

Step 3 Plot the points from the table of values. A line through these points is the graph of the equation.

Example 3 ***Graph a Linear Equation***

Use a table of values to graph the equation $x + 4y = 4$.

1. Rewrite the equation in function form by solving for y.

$x + 4y = 4$ — **Write original equation.**

$4y = -x + 4$ — **Subtract x from each side.**

$y = -\frac{1}{4}x + 1$ — **Divide each side by 4.**

2. Choose a few values of x and make a table of values.

When graphing a linear equation, try choosing values of x that include negative values, zero, and positive values to see how the graph behaves to the left and right of the y-axis.

x	−4	0	4
y	2	1	0

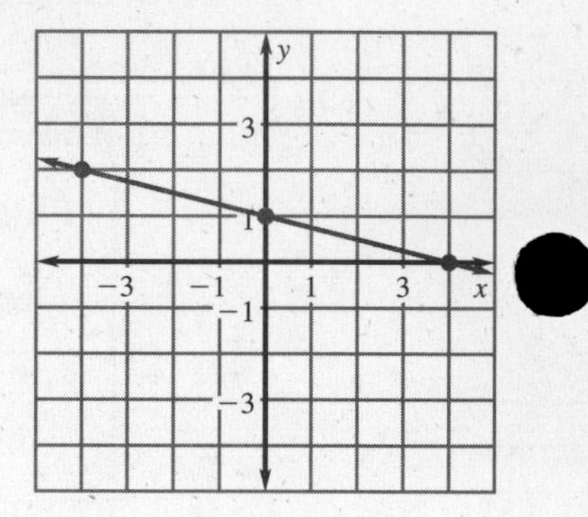

You have found three solutions.

(−4, 2), (0, 1), (4, 0)

3. Plot the points and draw a line through them.

✔ ***Checkpoint*** **Complete the following exercise.**

4. Use a table of values to graph the equation $x - 2y = 1$.

4.3 Graphing Horizontal and Vertical Lines

Goal Graph horizontal and vertical lines.

VOCABULARY

Constant function A constant function is a function of the form $y = b$, where b is a number.

EQUATIONS OF HORIZONTAL AND VERTICAL LINES

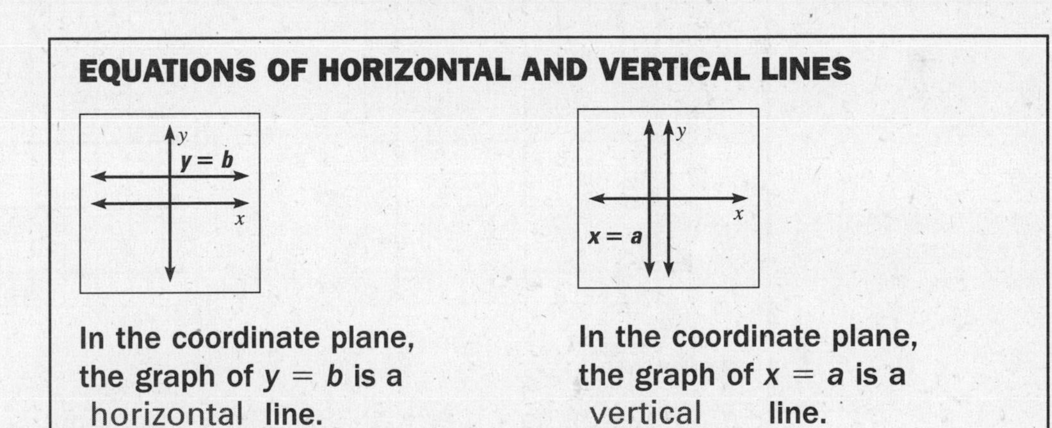

In the coordinate plane, the graph of $y = b$ is a horizontal line.

In the coordinate plane, the graph of $x = a$ is a vertical line.

Example 1 ***Graph the Equation $y = b$***

Graph the equation $y = -3$.

The equation does not have x as a variable. The y-coordinate is always -3, regardless of the value of x. Some points that are solutions of the equation are:

$(-3, -3)$, $(0, -3)$, and $(3, -3)$

The graph of $y = -3$ is a horizontal line 3 units below the x-axis.

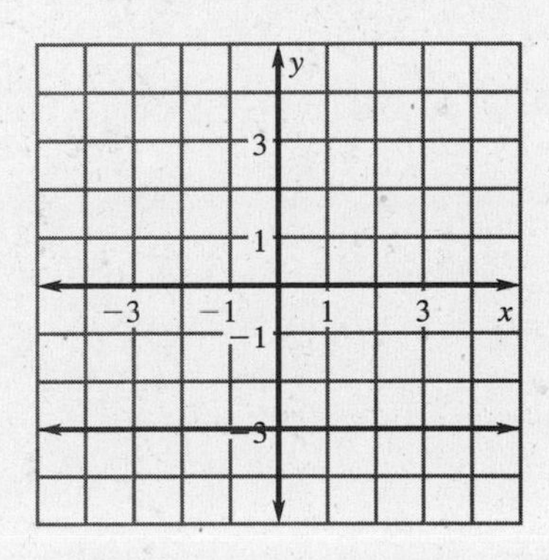

Example 2 *Graph the Equation $x = a$*

Graph the equation $x = 2$.

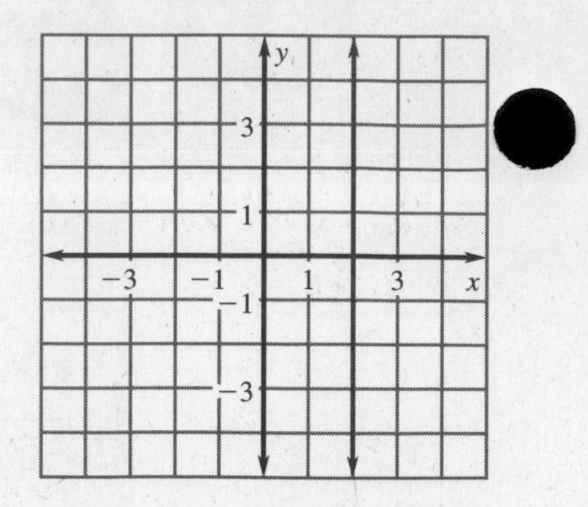

Solution

The equation does not have y as a variable. The x-coordinate is always 2, regardless of the value of y. Some points that are solutions of the equation are:

(2, −3), (2, 0), and (2, 3)

Answer The graph of $x = 2$ is a vertical line 2 units to the right of the y-axis.

Example 3 *Write an Equation of a Line*

Write the equation of the line in the graph.

a. 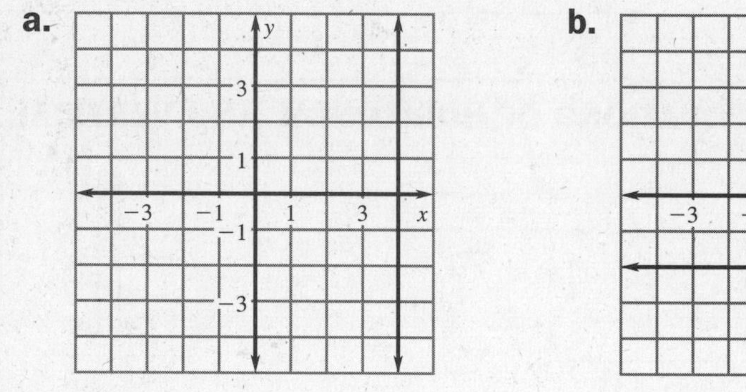

b. 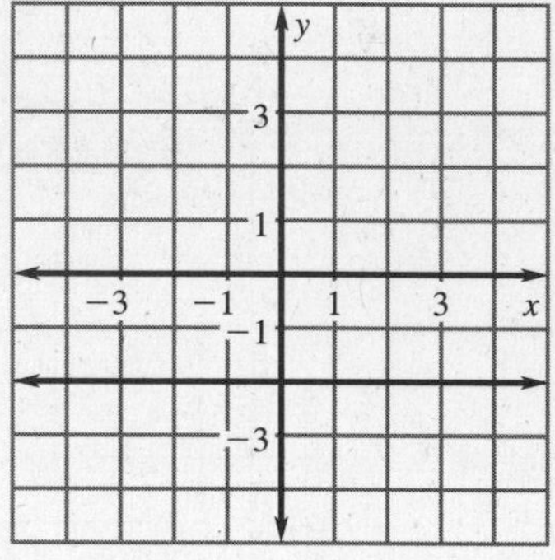

Solution

a. The graph is a vertical line. The x-coordinate is always 4. The equation of the line is $x = 4$.

b. The graph is a horizontal line. The y-coordinate is always -2. The equation of the line is $y = -2$.

 ***Checkpoint*** **Complete the following exercises.**

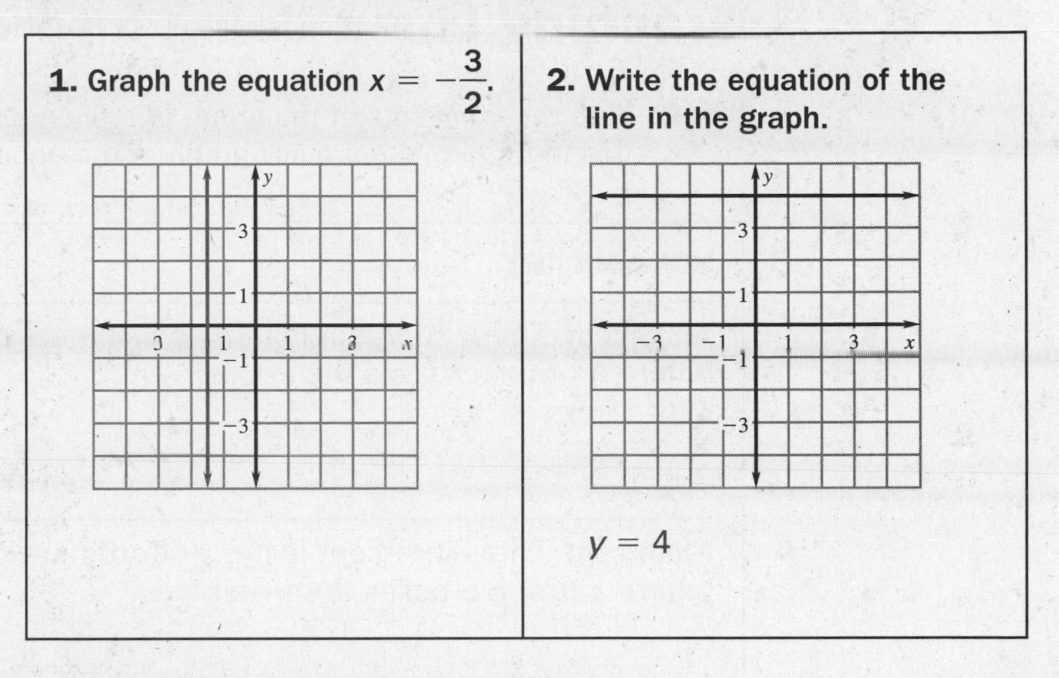

Example 4 ***Write a Constant Function***

Tree Trunks The graph shows the diameter of a tree trunk over a 6-week period. Write an equation to represent the diameter of the tree trunk for this period. What is the domain of the function? What is the range?

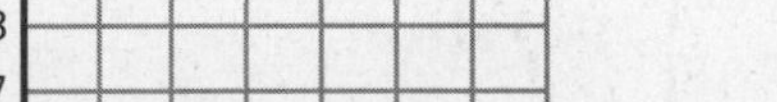

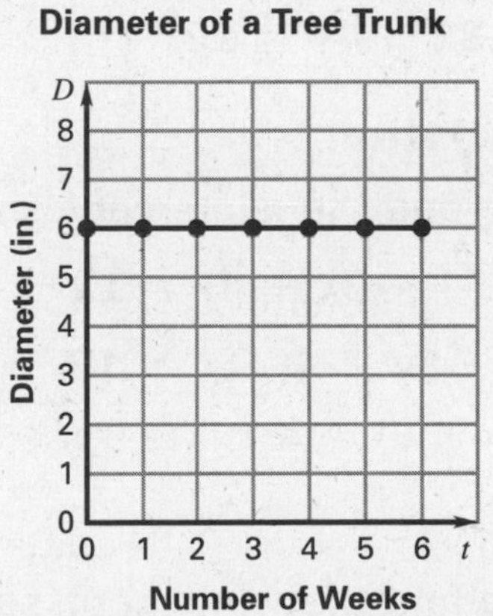

Solution

From the graph, you can see that the diameter was about 6 inches throughout the 6-week period. Therefore, the diameter D during this time t is $D =$ 6. The domain is $0 \le t \le 6$. The range is the single number 6.

4.4 Graphing Lines Using Intercepts

Goal Find the intercepts of the graph of a linear equation and then use them to make a quick graph of the equation.

VOCABULARY

***x*-intercept** The x-intercept is the x-coordinate of a point where a graph crosses the x-axis.

***y*-intercept** The y-intercept is the y-coordinate of a point where a graph crosses the y-axis.

Example 1 ***Find x- and y-Intercepts***

Find the x- and y-intercepts of the graph of the equation $-3x + 4y = 12$.

Solution

To find the x-intercept, substitute 0 for y and solve for x.

$-3x + 4y = 12$	**Write original equation.**
$-3x + 4(0) = 12$	**Substitute 0 for y.**
$-3x = 12$	**Simplify.**
$x = -4$	**Solve for x.**

To find the y-intercept, substitute 0 for x and solve for y.

$-3x + 4y = 12$	**Write original equation.**
$-3(0) + 4y = 12$	**Substitute 0 for x.**
$4y = 12$	**Simplify.**
$y = 3$	**Solve for y.**

Answer The x-intercept is -4. The y-intercept is 3.

✓ *Checkpoint* **Complete the following exercise.**

1. Find the x-intercept and the y-intercept of the graph of the equation $2x - 5y = 10$.

x-intercept: 5; y-intercept: -2

The Quick Graph process works because only two points are needed to determine a line.

MAKING A QUICK GRAPH

Step 1 **Find the** intercepts **.**

Step 2 **Draw a coordinate plane that includes the** intercepts **.**

Step 3 **Plot the** intercepts **and draw a line through them.**

Example 2 ***Make a Quick Graph***

Graph the equation $9x + 6y = 18$.

Solution

1. Find the intercepts.

$9x + 6y = 18$	**Write original equation.**
$9x + 6(\underline{0}) = 18$	**Substitute** 0 **for** ***y*****.**
$x = \underline{2}$	**The *x*-intercept is** 2 **.**
$9x + 6y = 18$	**Write original equation.**
$9(\underline{0}) + 6y = 18$	**Substitute** 0 **for** ***x*****.**
$y = \underline{3}$	**The *y*-intercept is** 3 **.**

2. Draw a coordinate plane that includes the points (2 **,** 0 **) and (** 0 **,** 3 **).**

3. Plot the points (2 **,** 0 **) and (** 0 **,** 3 **) and draw a line through them.**

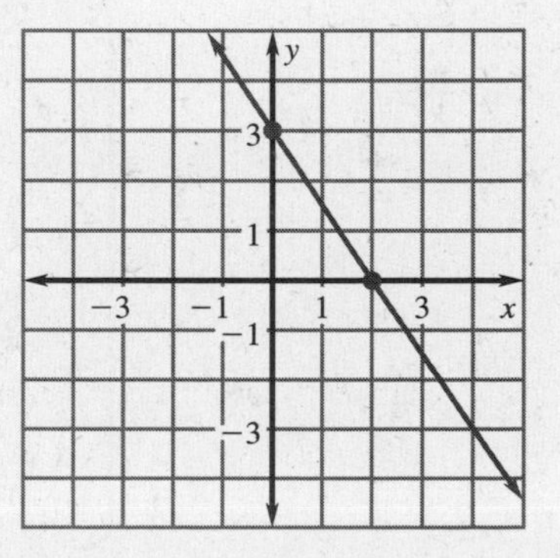

 ***Checkpoint*** **Complete the following exercise.**

2. Graph the equation $-4x + 5y = 20$.

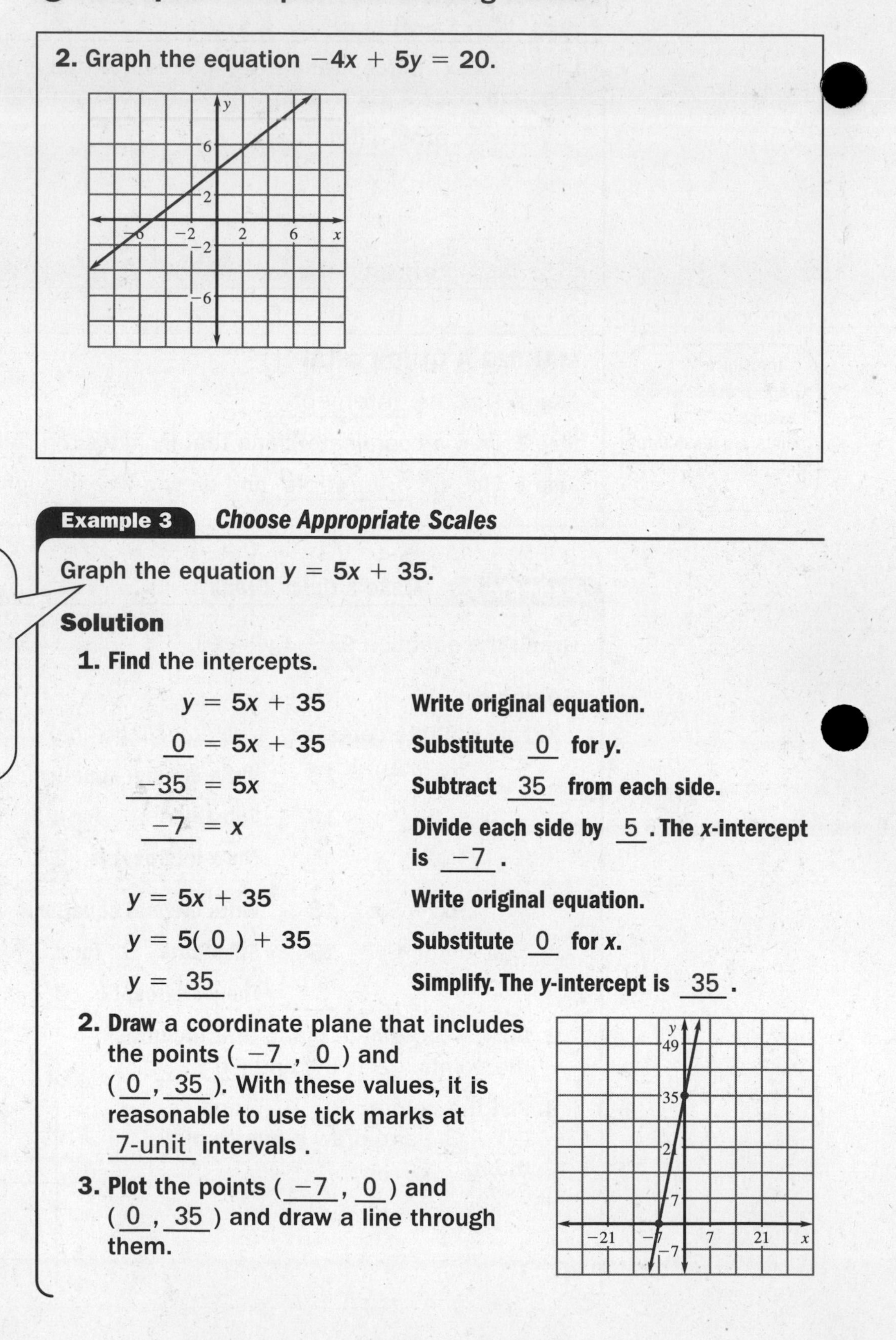

Example 3 ***Choose Appropriate Scales***

When you make a quick graph, find the intercepts *before* you draw the coordinate plane. This will help you find an appropriate scale on each axis.

Graph the equation $y = 5x + 35$.

Solution

1. Find the intercepts.

$y = 5x + 35$	**Write original equation.**
$\underline{0} = 5x + 35$	**Substitute** $\underline{0}$ **for** y.
$\underline{-35} = 5x$	**Subtract** $\underline{35}$ **from each side.**
$\underline{-7} = x$	**Divide each side by** $\underline{5}$. **The** x**-intercept is** $\underline{-7}$.
$y = 5x + 35$	**Write original equation.**
$y = 5(\underline{0}) + 35$	**Substitute** $\underline{0}$ **for** x.
$y = \underline{35}$	**Simplify. The** y**-intercept is** $\underline{35}$.

2. Draw a coordinate plane that includes the points ($\underline{-7}$, $\underline{0}$) and ($\underline{0}$, $\underline{35}$). With these values, it is reasonable to use tick marks at $\underline{\text{7-unit}}$ intervals .

3. Plot the points ($\underline{-7}$, $\underline{0}$) and ($\underline{0}$, $\underline{35}$) and draw a line through them.

4.5 The Slope of a Line

Goal Find the slope of a line.

VOCABULARY

Slope The slope of a line is the ratio of the vertical rise to the horizontal run between any two points on the line.

Example 1 ***The Slope Ratio***

Find the slope of a ramp that has a vertical rise of 3 feet and a horizontal run of 18 feet. Let m represent the slope.

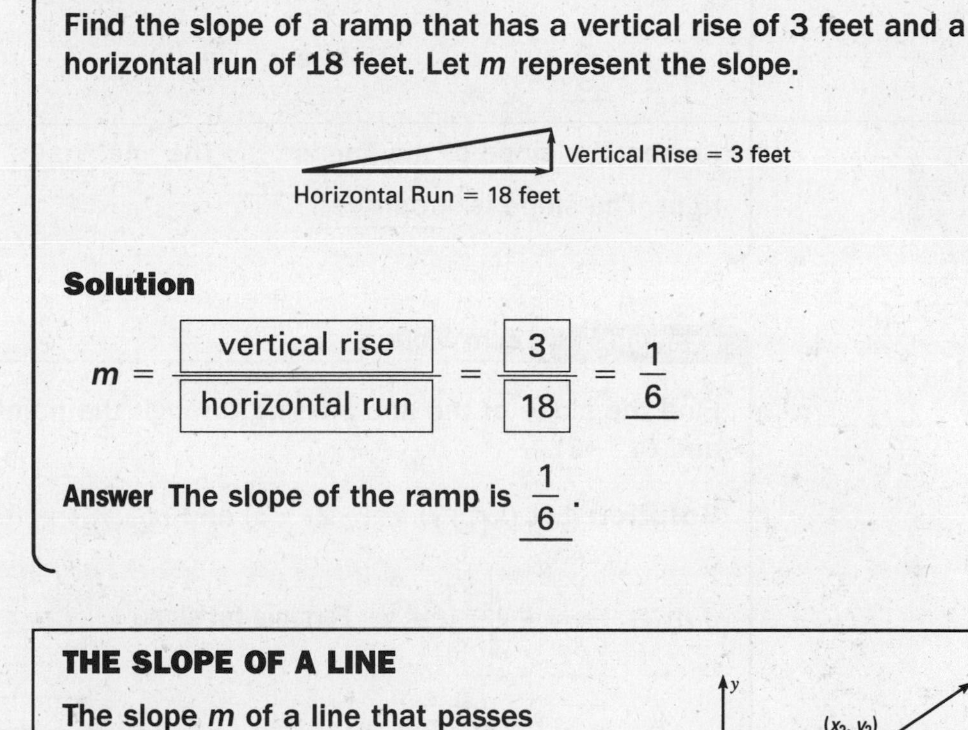

Solution

$$m = \frac{\boxed{\text{vertical rise}}}{\boxed{\text{horizontal run}}} = \frac{\boxed{3}}{\boxed{18}} = \frac{1}{6}$$

Answer The slope of the ramp is $\frac{1}{6}$.

THE SLOPE OF A LINE

The slope m of a line that passes through the points (x_1, y_1) and (x_2, y_2) is

$$m = \frac{\text{rise}}{\text{run}} = \frac{\text{change in } \boxed{y}}{\text{change in } \boxed{x}}$$

$$= \frac{\boxed{y_2 - y_1}}{\boxed{x_2 - x_1}}.$$

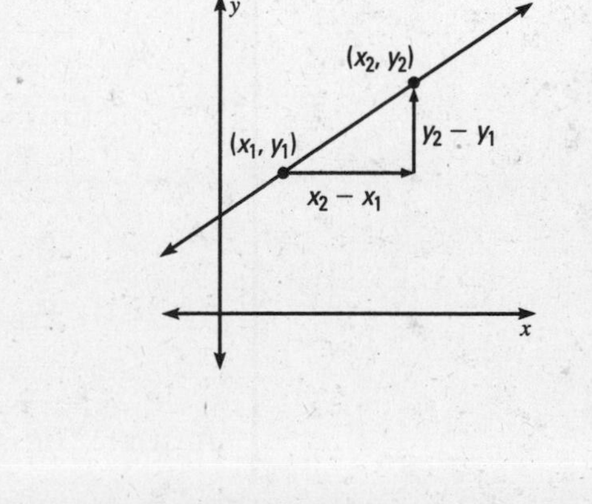

In the slope formula, x_1 is read as "x sub one" and y_2 as "y sub two."

Example 2 *Positive Slope*

Find the slope of the line that passes through the points (1, 2) and (−2, −3).

Solution Let $(x_1, y_1) = (1, 2)$ and $(x_2, y_2) = (\ \ 2,\ \ 3)$.

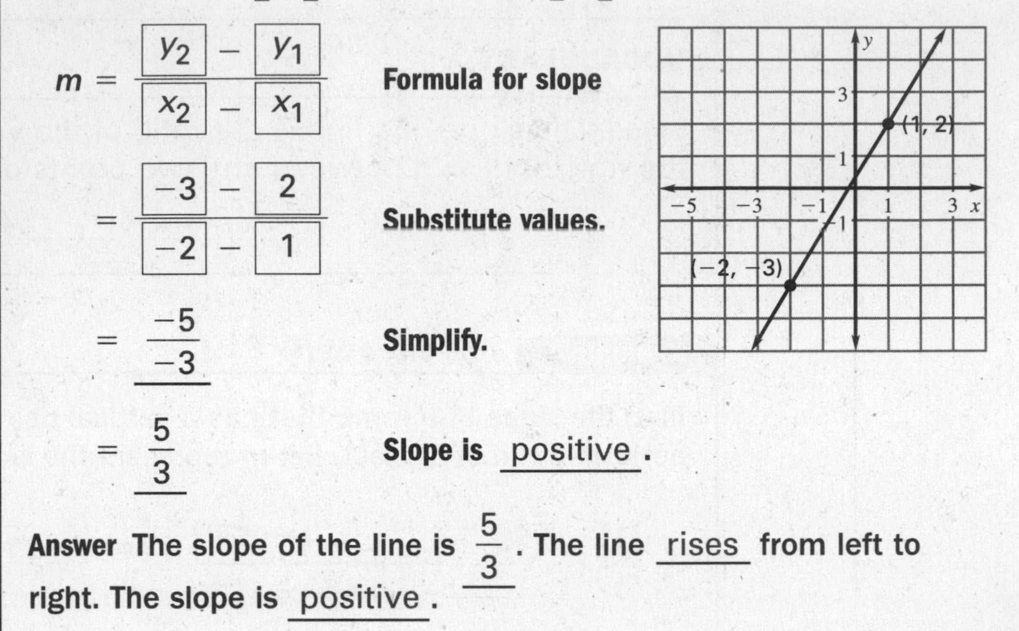

$$m = \frac{y_2 - y_1}{x_2 - x_1}$$ **Formula for slope**

$$= \frac{-3 - 2}{-2 - 1}$$ **Substitute values.**

$$= \frac{-5}{-3}$$ **Simplify.**

$$= \frac{5}{3}$$ **Slope is** positive.

Answer The slope of the line is $\frac{5}{3}$. The line rises from left to right. The slope is positive.

Example 3 *Zero Slope*

Find the slope of the line passing through the points (−2, −3) and (4, −3).

Solution Let $(x_1, y_1) = (-2, -3)$ and $(x_2, y_2) = (4, -3)$.

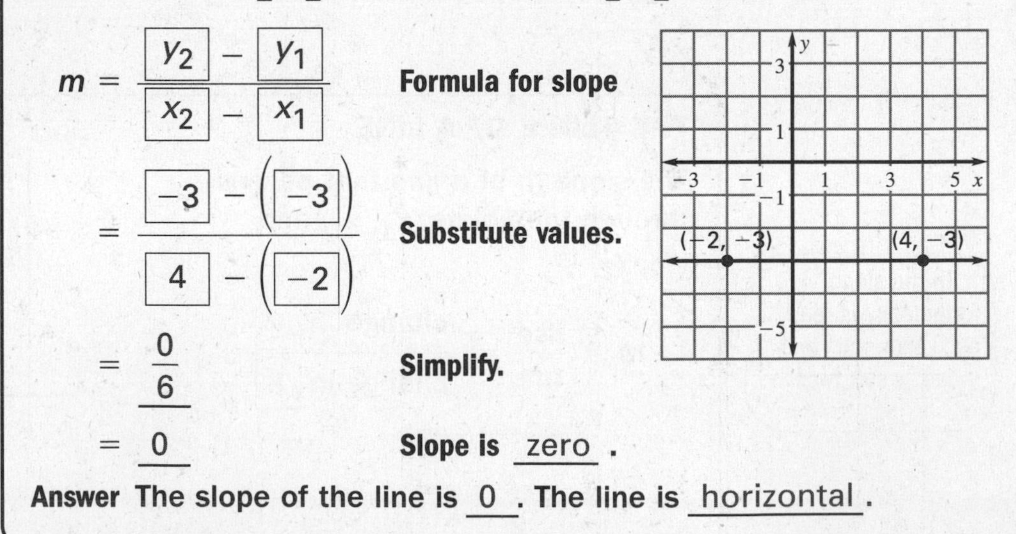

$$m = \frac{y_2 - y_1}{x_2 - x_1}$$ **Formula for slope**

$$= \frac{-3 - (-3)}{4 - (-2)}$$ **Substitute values.**

$$= \frac{0}{6}$$ **Simplify.**

$$= 0$$ **Slope is** zero.

Answer The slope of the line is 0. The line is horizontal.

Example 4 *Undefined Slope*

Find the slope of the line passing through the points (−1, −4) and (−1, −2).

Solution

Let $(x_1, y_1) = (-1, -4)$ and $(x_2, y_2) = (-1, -2)$.

$$m = \frac{y_2 - y_1}{x_2 - x_1}$$ **Formula for slope**

$$= \frac{-2 - (-4)}{-1 - (-1)}$$ **Substitute values.**

$$= \frac{2}{0}$$ **Division by** 0 **is** undefined **.**

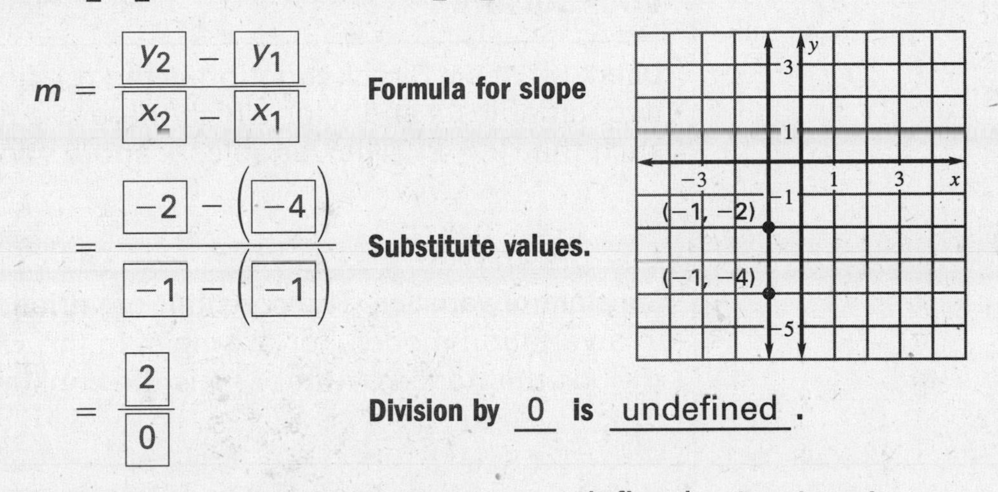

Answer Because division by zero is undefined, the slope is undefined. The line is vertical.

✔ ***Checkpoint*** **Find the slope of the line passing through the points. Then state whether the slope of the line is *positive*, *negative*, *zero*, or *undefined*.**

1. (−5, 2), (7, −2)	**2.** (0, 0), (−9, 0)
$-\frac{1}{3}$; negative slope	0; zero slope
3. (−7, −8), (−7, 8)	**4.** (2, −4), (8, 6)
undefined; undefined slope	$\frac{5}{3}$; positive slope

4.6 Direct Variation

Goal Write and graph equations that represent direct variation.

VOCABULARY

Direct variation Direct variation is the relationship between two variables x and y for which there is a nonzero number k such that $y = kx$. The variables x and y vary directly.

Constant of variation The constant of variation is the constant in a variation model. For example, in the variation model $y = kx$, the nonzero number k is the constant of variation.

Example 1 ***Write a Direct Variation Model***

The model for direct variation $y = kx$ is read as "y varies directly with x."

The variables x and y vary directly. One pair of values is $x = 7$ and $y = 21$.

a. Write an equation that relates x and y.

b. Find the value of y when $x = 4$.

Solution

a. Because x and y vary directly, the equation is of the form $y = kx$.

$y = kx$	**Write model for direct variation.**
$21 = k(7)$	**Substitute 7 for x and 21 for y.**
$3 = k$	**Divide each side by 7.**

Answer An equation that relates x and y is $y = 3x$.

b.

$y = 3(4)$	**Substitute 4 for x.**
$y = 12$	**Simplify.**

Answer When $x = 4$, $y = 12$.

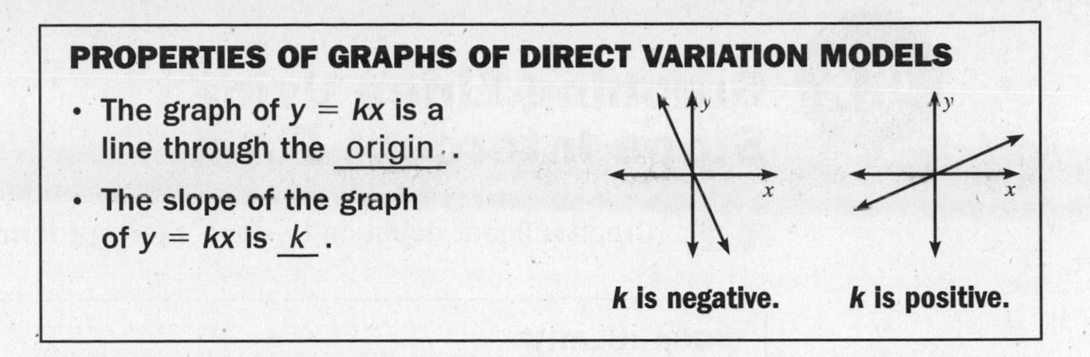

PROPERTIES OF GRAPHS OF DIRECT VARIATION MODELS

- The graph of $y = kx$ is a line through the origin.
- The slope of the graph of $y = kx$ is k.

Example 2 ***Graph a Direct Variation Model***

Graph the equation $y = -x$.

1. **Plot a point at the** origin.
2. **Find a second point by choosing any value of x and substituting it into the equation to find the corresponding y-value. Use the value -3 for x.**

$y = -x$	**Write original equation.**
$y = -(-3)$	**Substitute -3 for x.**
$y = 3$	**Simplify.**

   The second point is (-3, 3).

3. **Plot the second point and draw a line through the** origin **and the second point.**

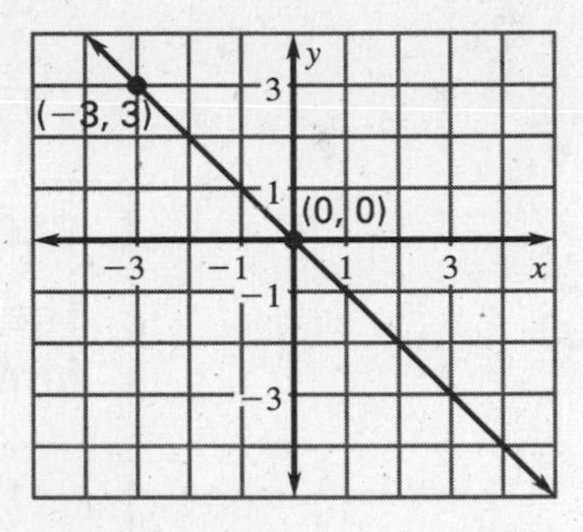

Checkpoint **The variables x and y vary directly. Use the given values to write an equation that relates x and y.**

1. $x = 6, y = 30$	**2.** $x = 8, y = -20$	**3.** $x = 3.6, y = 1.8$
$y = 5x$	$y = -\frac{5}{2}x$	$y = 0.5x$

4.7 Graphing Lines Using Slope-Intercept Form

Goal Graph a linear equation in slope-intercept form.

VOCABULARY

Slope-intercept form The slope-intercept form of a linear equation is written in the form $y = mx + b$. The slope of the line is m and the y-intercept is b.

Parallel lines Parallel lines are two different lines in the same plane that do not intersect.

SLOPE-INTERCEPT FORM OF THE EQUATION OF A LINE

The linear equation $y = mx + b$ is written in **slope-intercept form,** where m is the slope and b is the y-intercept.

slope ↓ y-intercept ↓

$$y = mx + b$$

Example 1 ***Find the Slope and y-Intercept***

Find the slope and y-intercept of $-3x - y = 2$.

Solution Rewrite the equation in slope-intercept form.

$-3x - y = 2$	**Write original equation.**
$-y = 3x + 2$	**Add $3x$ to each side.**
$y = -3x - 2$	**Divide each side by -1.**
	$m = -3$ **and** $b = -2$.

Answer The slope is -3. The y-intercept is -2.

Example 2 *Graph an Equation in Slope-Intercept Form*

Graph the equation $y = 2x - 3$.

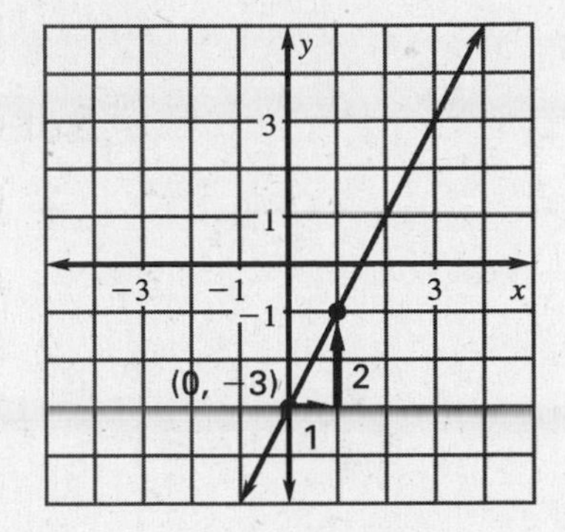

1. **Find the slope,** 2 **, and the y-intercept,** −3 **.**
2. **Plot the point $(0, b)$ when b is** −3 **.**
3. **Use the slope to locate a second point on the line.**

$$m = \frac{\boxed{2}}{\boxed{1}} = \frac{\text{rise}}{\text{run}} \rightarrow \frac{\text{move } \boxed{2} \text{ units up}}{\text{move } \boxed{1} \text{ unit right}}$$

4. **Draw a line through the two points.**

Checkpoint **Find the slope and y-intercept of the equation.**

1. $y = 4 - 3x$	**2.** $2x + y = -3$	**3.** $4y = 3x - 8$
-3; 4	-2; -3	$\frac{3}{4}$; -2

Graph the equation in slope-intercept form.

4. $y = x - 2$

5. $y = -\frac{1}{4}x + 1$

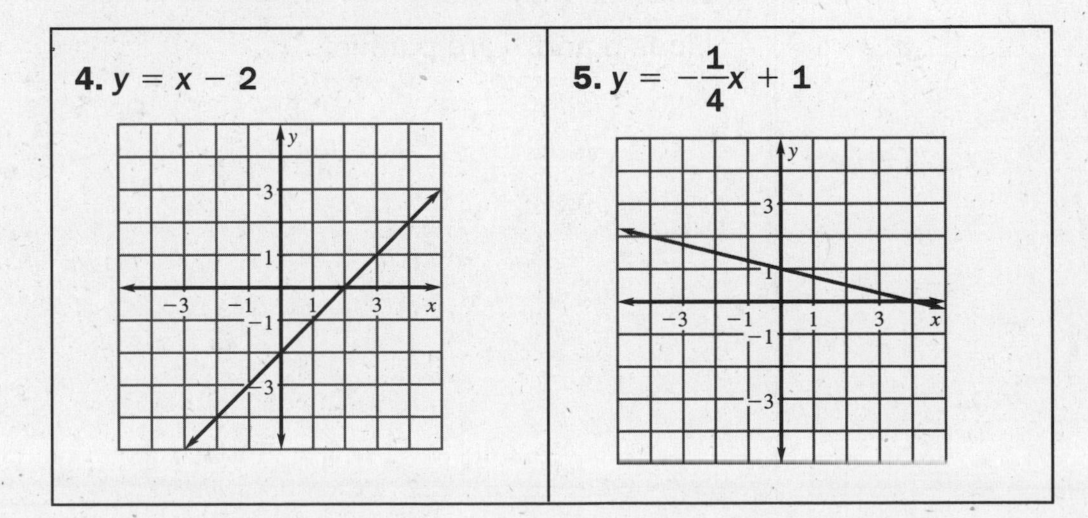

Example 3 Identify Parallel Lines

Which of the following lines are parallel?

line *a*: $-2x + y = 1$ **line *b*:** $2x + y = -1$ **line *c*:** $2x - y = 3$

Solution

1. Rewrite each equation in slope-intercept form.

line *a*: $y =$ $2x + 1$ **line *b*:** $y =$ $-2x - 1$ **line *c*:** $y =$ $2x - 3$

2. Identify the slope of each equation.

The slope of line *a* is 2. The slope of line *b* is -2. The slope of line *c* is 2.

3. Compare the slopes.

Lines *a* and *c* are parallel because each has a slope of 2.

Line *b* is *not* parallel to either of the other two lines because it has a slope of -2.

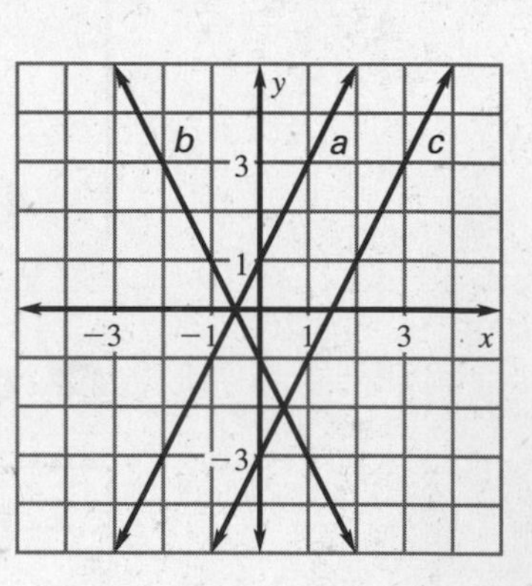

Check The graph gives you a visual check. It shows that line *b* intersects each of the two parallel lines.

Answer Lines *a* and *c* are parallel.

Checkpoint **Which of the following lines are parallel?**

6. line *a*: $4x - 3y = 6$
line *b*: $-8x + 6y = 18$
line *c*: $4x + 3y = 8$

Lines *a* and *b* are parallel.

4.8 Functions and Relations

Goal Decide whether a relation is a function and use function notation.

VOCABULARY

Relation A relation is any set of ordered pairs.

Function A function is a rule that establishes a relationship between two quantities, called the input and output. There is exactly one output for each input.

Function notation Function notation is a way to describe a function by means of an equation. For the equation $y = f(x)$, the symbol $f(x)$ denotes the output and is read as "the value of f at x" or simply as "f of x."

Linear function A function is a linear function if it is of the form $f(x) = mx + b$.

Example 1 ***Identify Functions***

Decide whether the relation is a function. If it is a function, give the domain and the range.

a.

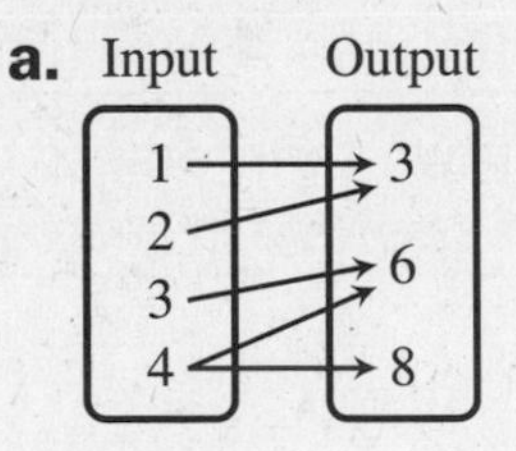

b.

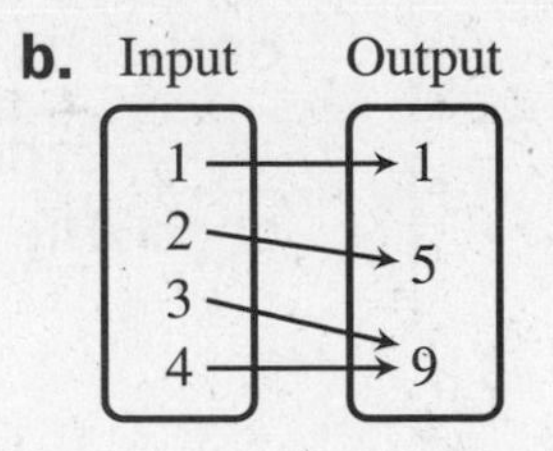

Solution

a. **The relation** is not **a function because** the input 4 has two outputs: 6 and 8.

b. **The relation** is **a function. For each** input **there is** exactly one output. **The domain of the function is** 1, 2, 3, and 4. **The range is** 1, 5, and 9.

VERTICAL LINE TEST FOR FUNCTIONS

A graph is a function if no vertical line intersects the graph at more than one point.

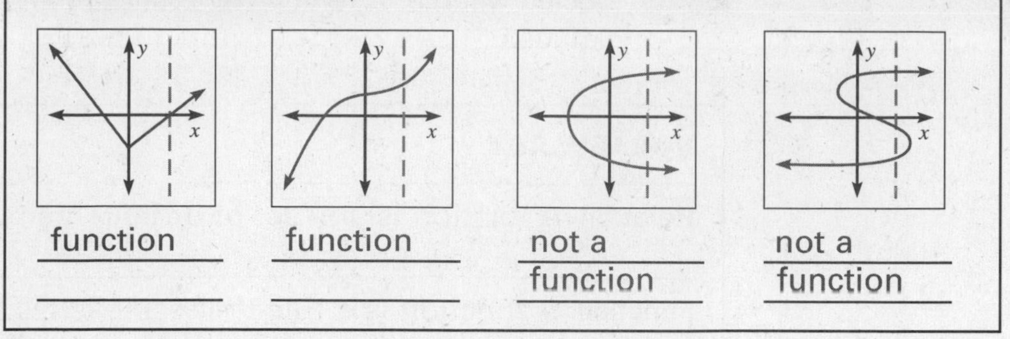

function | function | not a function | not a function

Example 2 *Use the Vertical Line Test*

Use the vertical line test to determine whether the graph represents a function.

a. **b.**

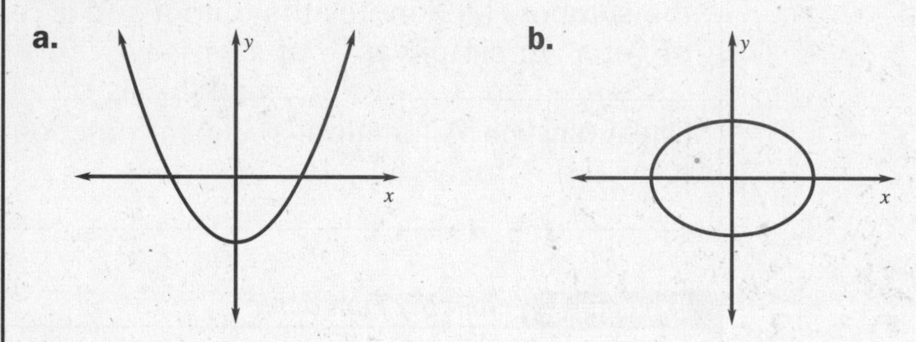

Solution

a. No vertical line can be drawn to intersect the graph more than once. The graph represents y as a function of x.

b. A vertical line can be drawn to intersect the graph more than once. The graph does not represent y as a function of x.

✓ *Checkpoint* **Decide whether the relation is a function. If it is a function, give the domain and the range.**

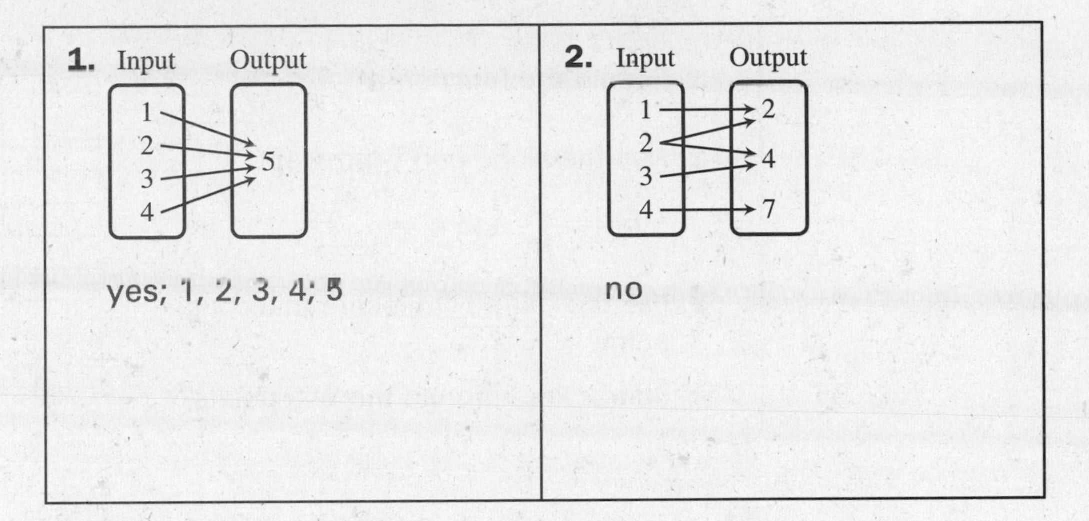

Use the vertical line test to determine whether the graph represents a function.

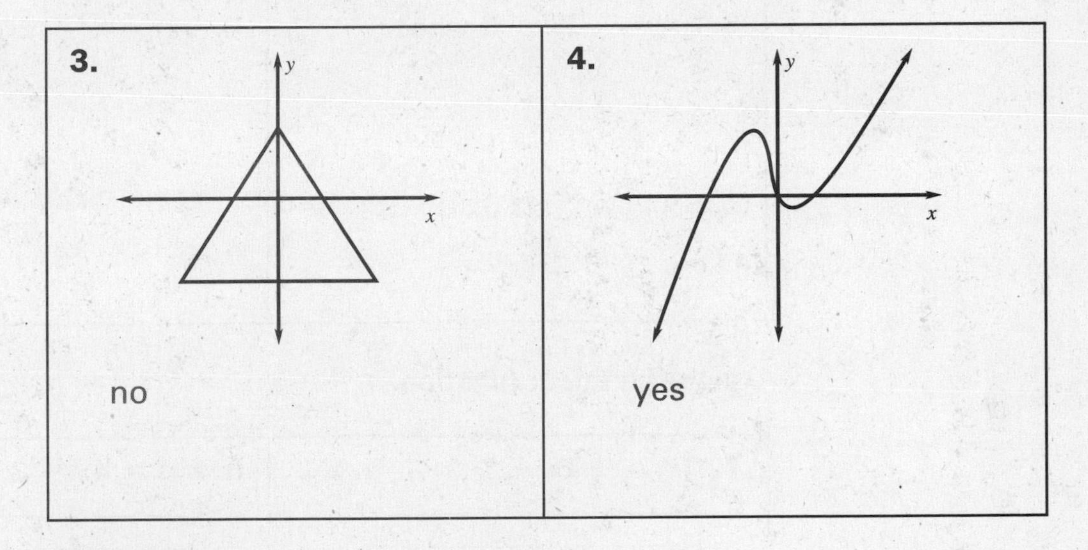

Example 3 ***Evaluate a Function***

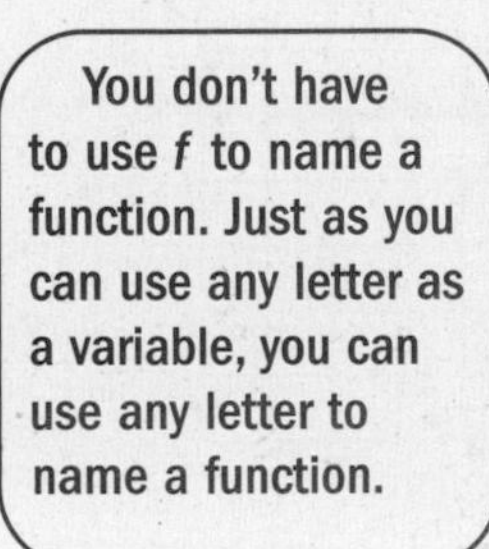

Evaluate $g(x) = -2x + 3$ when $x = 4$.

Solution

$g(x) = -2x + 3$	**Write original function.**
$g(\underline{4}) = -2(\underline{4}) + 3$	**Substitute** $\underline{4}$ **for** x.
$= \underline{-5}$	**Simplify.**

Answer When $x = 4$, $g(x) = \underline{-5}$.

Example 4 *Graph a Linear Function*

Graph $f(x) = \frac{3}{4}x - 2$.

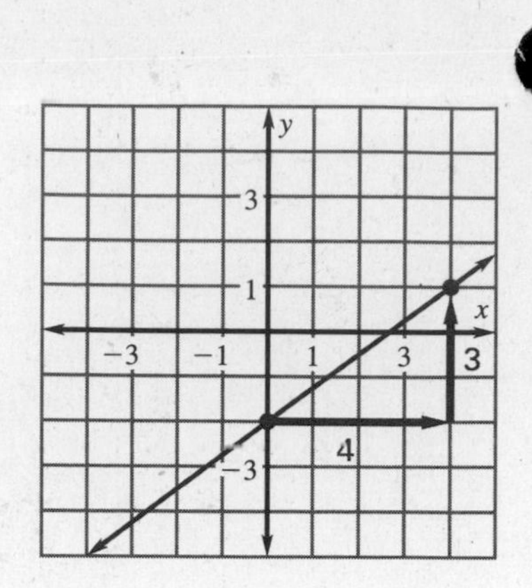

1. Rewrite the function as $y = \underline{\frac{3}{4}x - 2}$.

2. Find the slope and *y*-intercept.

$m = \underline{\frac{3}{4}}$ and $b = \underline{-2}$

3. Use the <u>slope</u> **to locate a second point.**

4. Draw a line through the two points.

✔ ***Checkpoint*** **Evaluate the function for the given value of the variable.**

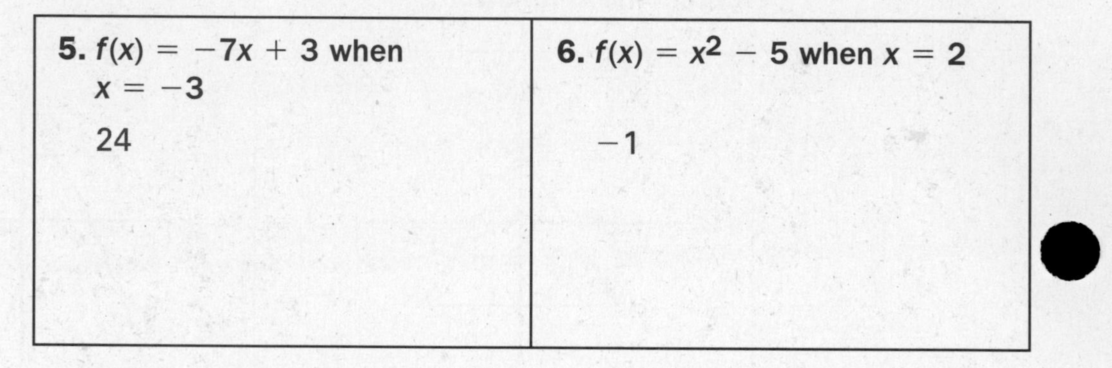

5. $f(x) = -7x + 3$ when $x = -3$

24

6. $f(x) = x^2 - 5$ when $x = 2$

−1

Graph the linear function.

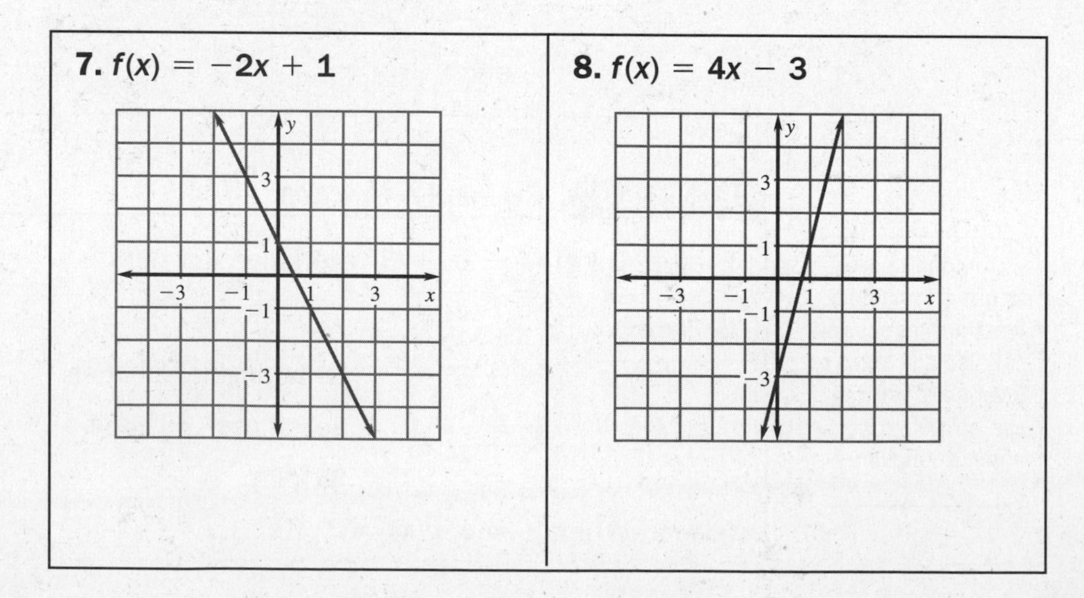

7. $f(x) = -2x + 1$

8. $f(x) = 4x - 3$

Words to Review

Give an example of the vocabulary word.

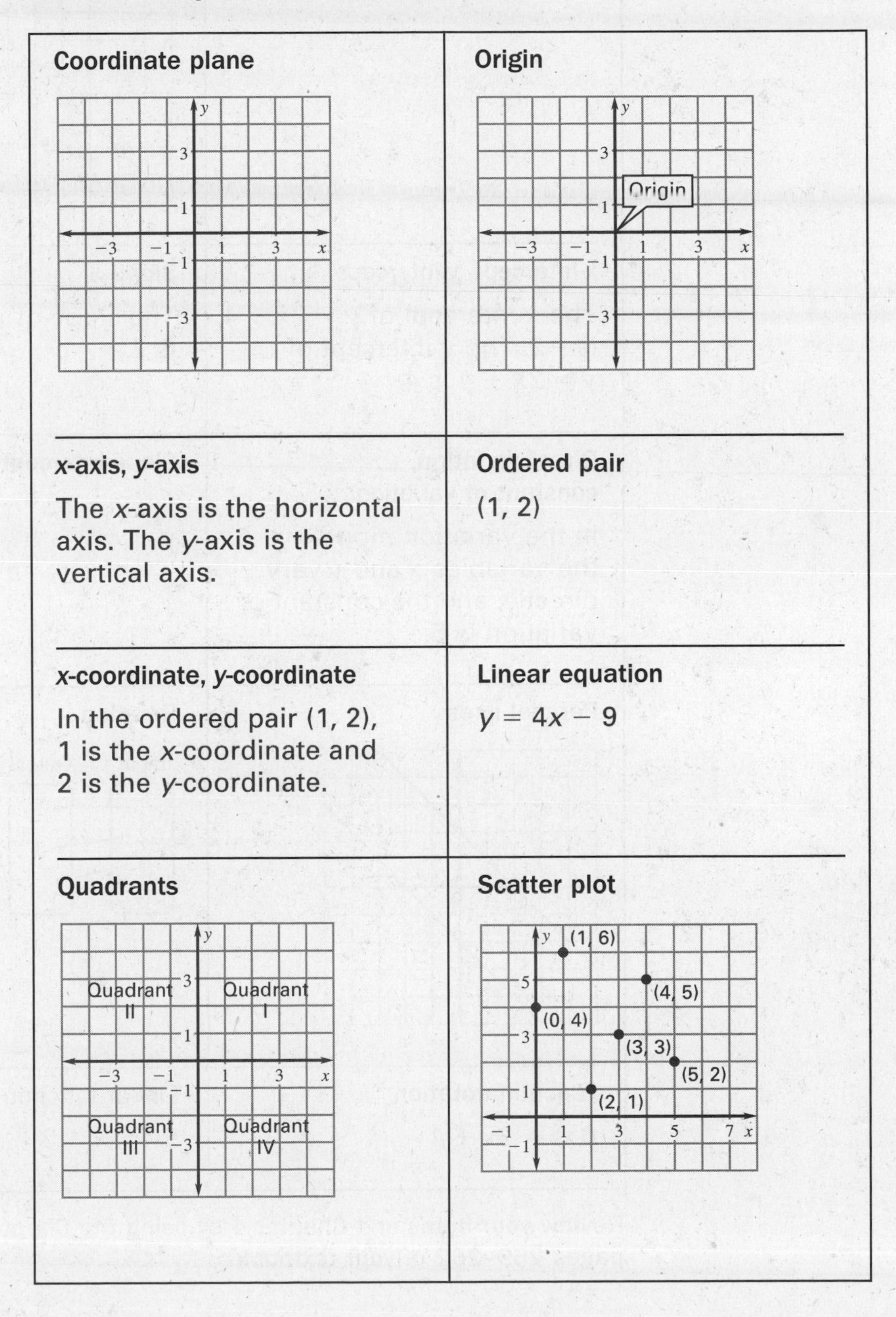

Coordinate plane

Origin

x-axis, y-axis

The x-axis is the horizontal axis. The y-axis is the vertical axis.

Ordered pair

(1, 2)

x-coordinate, y-coordinate

In the ordered pair (1, 2), 1 is the x-coordinate and 2 is the y-coordinate.

Linear equation

$y = 4x - 9$

Quadrants

Scatter plot

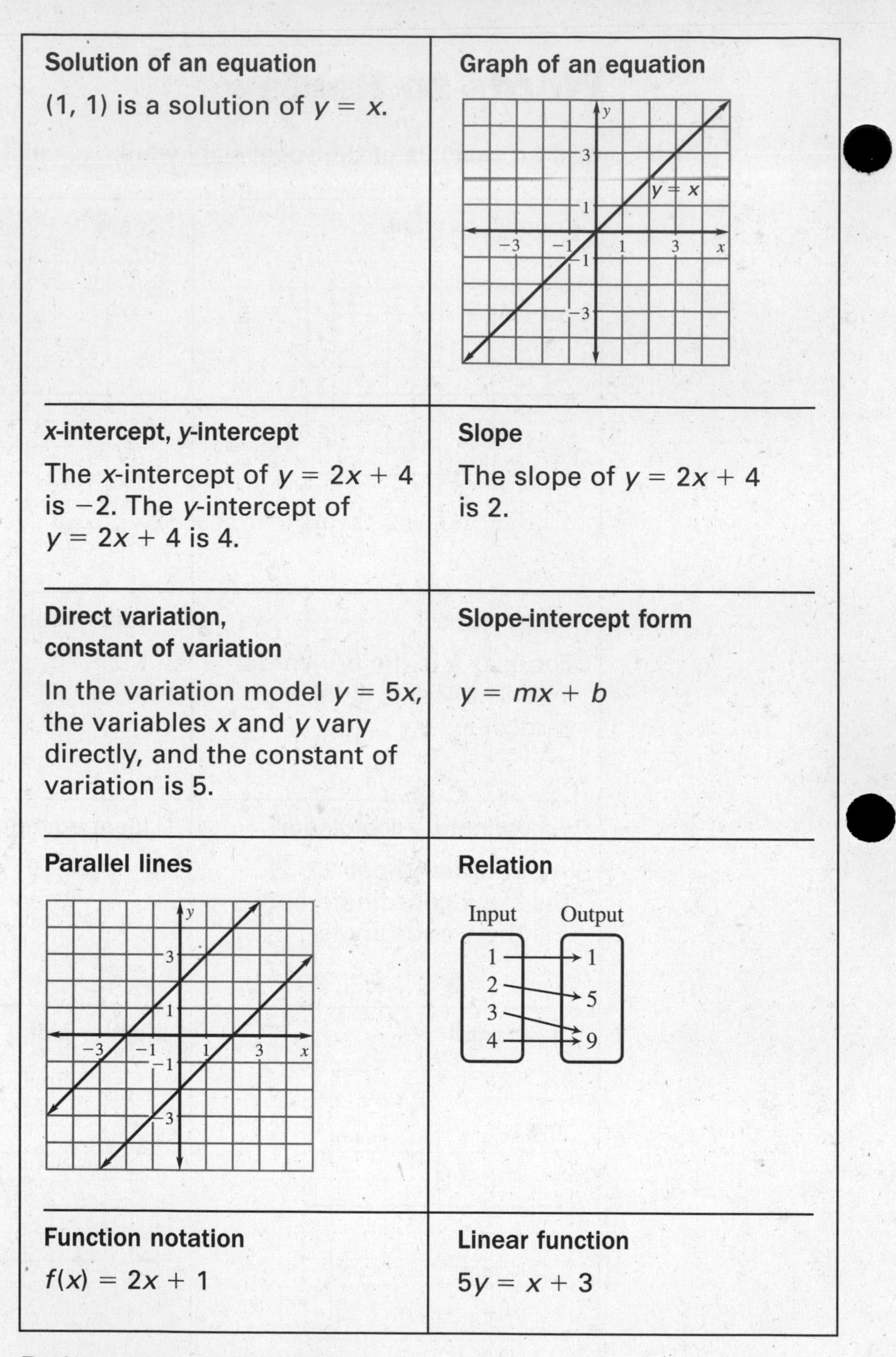

Solution of an equation (1, 1) is a solution of $y = x$.	**Graph of an equation**
x-intercept, y-intercept The x-intercept of $y = 2x + 4$ is -2. The y-intercept of $y = 2x + 4$ is 4.	**Slope** The slope of $y = 2x + 4$ is 2.
Direct variation, constant of variation In the variation model $y = 5x$, the variables x and y vary directly, and the constant of variation is 5.	**Slope-intercept form** $y = mx + b$
Parallel lines	**Relation**
Function notation $f(x) = 2x + 1$	**Linear function** $5y = x + 3$

Review your notes and Chapter 4 by using the Chapter Review on pages 259–262 of your textbook.

5.1 Slope-Intercept Form

Goal Use the slope-intercept form to write an equation of a line.

SLOPE-INTERCEPT FORM

The slope-intercept form of the equation of a line with slope m and y-intercept b is

$y - mx + b$

Example 1 ***Equation of a Line***

Write an equation of the line whose slope is 4 and whose y-intercept is -3.

Solution

1. **Write the slope-intercept form.** $y = mx + b$
2. **Substitute slope 4 for m and -3 for b.** $y = 4x + (-3)$
3. **Simplify the equation.** $y = 4x - 3$

Answer The equation of the line is $y = 4x - 3$.

Checkpoint **Write an equation of the line in slope-intercept form.**

1. The slope is 3 and the y-intercept is 7. $y = 3x + 7$	**2.** The slope is -5 and the y-intercept is 1. $y = -5x + 1$

Example 2 *Use a Graph to Write an Equation*

Write the equation of the line shown in the graph using slope-intercept form.

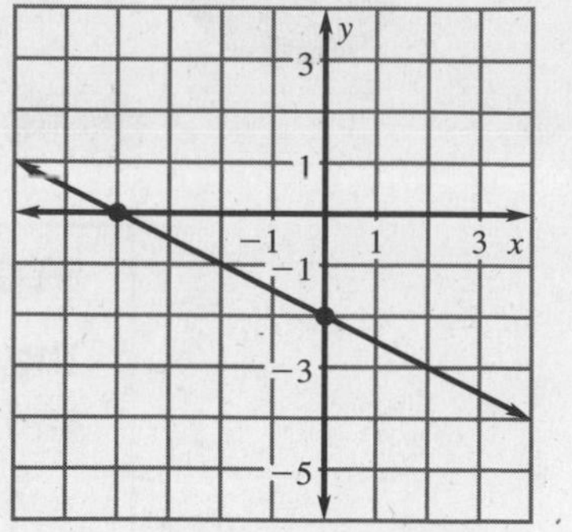

Solution

1. Write the slope-intercept form $y = mx + b$.

2. Find the slope m **of the line. Use any two points on the graph. Let** $(-4, 0)$ **be** (x_1, y_1) **and** $(0, -2)$ **be** (x_2, y_2).

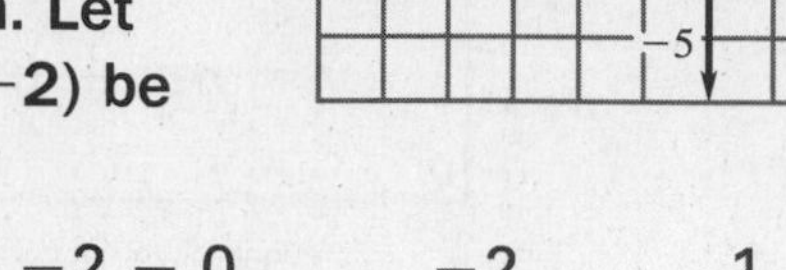

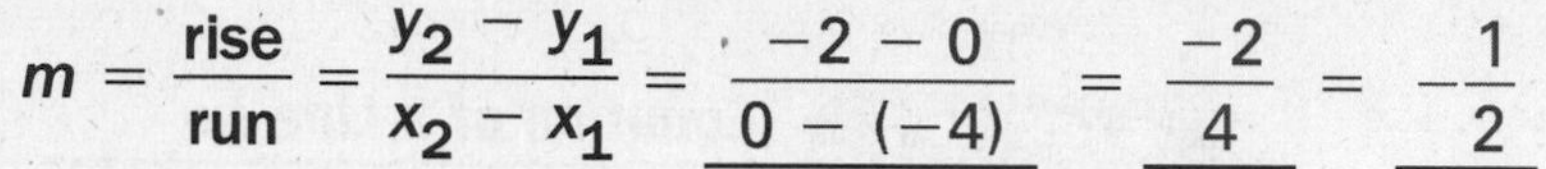

$$m = \frac{\text{rise}}{\text{run}} = \frac{y_2 - y_1}{x_2 - x_1} = \frac{-2 - 0}{0 - (-4)} = \frac{-2}{4} = -\frac{1}{2}$$

Recall that the *y*-intercept is the *y*-coordinate of the point where the line crosses the *y*-axis.

3. Use the graph to find the ***y*****-intercept** b**. The graph of the line crosses the** ***y*****-axis at (** 0 **,** −2 **). The** ***y*****-intercept is** −2 **.**

4. Substitute slope $-\frac{1}{2}$ **for** m **and** −2 **for** b **in the equation** $y = mx + b$.

$$y = -\frac{1}{2}x + (-2)$$

Answer The equation of the line is $y = -\frac{1}{2}x - 2$.

Checkpoint **Write an equation of the line in slope-intercept form.**

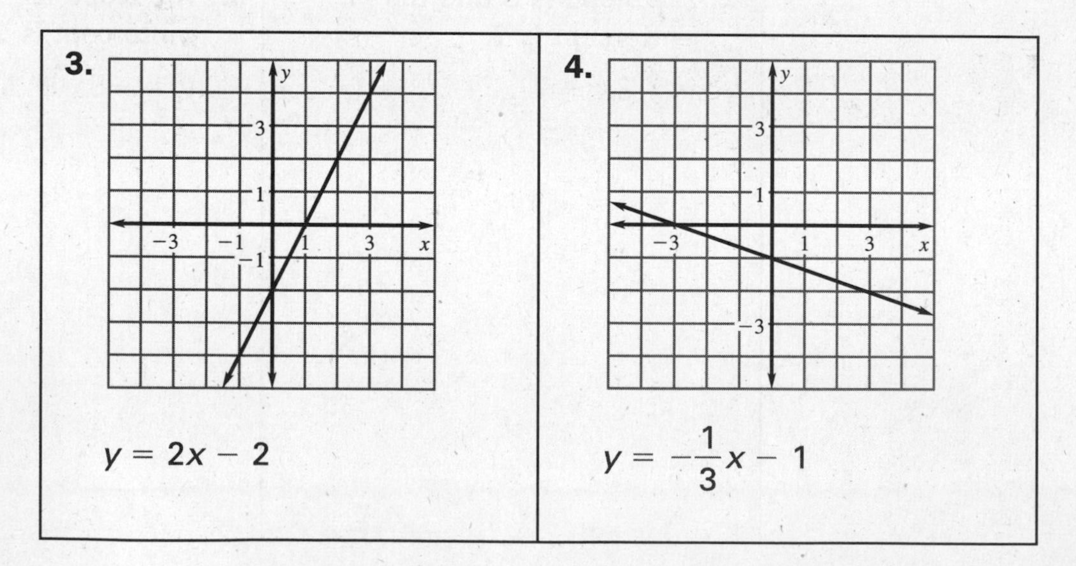

3. $y = 2x - 2$

4. $y = -\frac{1}{3}x - 1$

5.2 Point-Slope Form

Goal Use point-slope form to write the equation of a line.

POINT-SLOPE FORM

The point-slope form of the equation of the line through (x_1, y_1) with slope m is $y - y_1 = m(x - x_1)$.

Example 1 ***Point-Slope Form from a Graph***

Write the equation of the line in the graph in point-slope form.

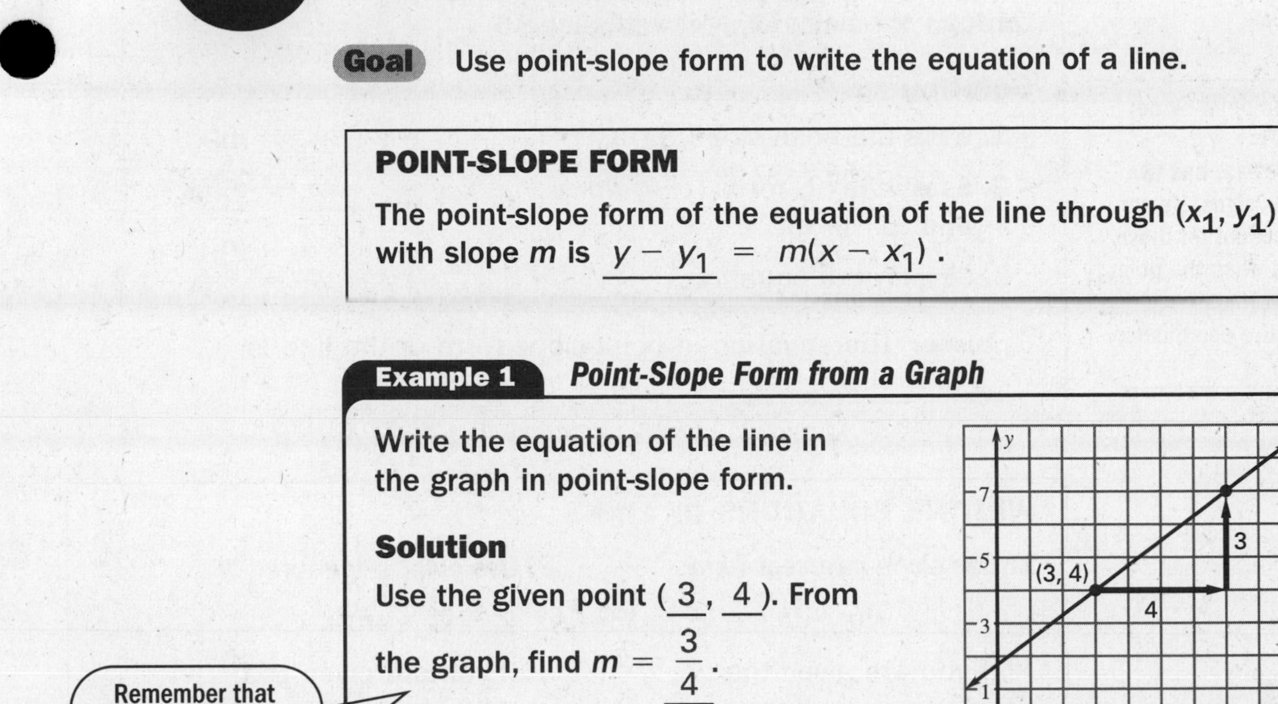

Solution

Use the given point (3 , 4). From the graph, find $m = \frac{3}{4}$.

> Remember that you can calculate slope as $m = \frac{\text{rise}}{\text{run}}$.

$y - y_1 = m(x - x_1)$ **Write point-slope form.**

$y - 4 = \frac{3}{4}(x - 3)$ **Substitute $\frac{3}{4}$ for m, 3 for x_1, and 4 for y_1.**

Checkpoint Write the equation of the line in point-slope form.

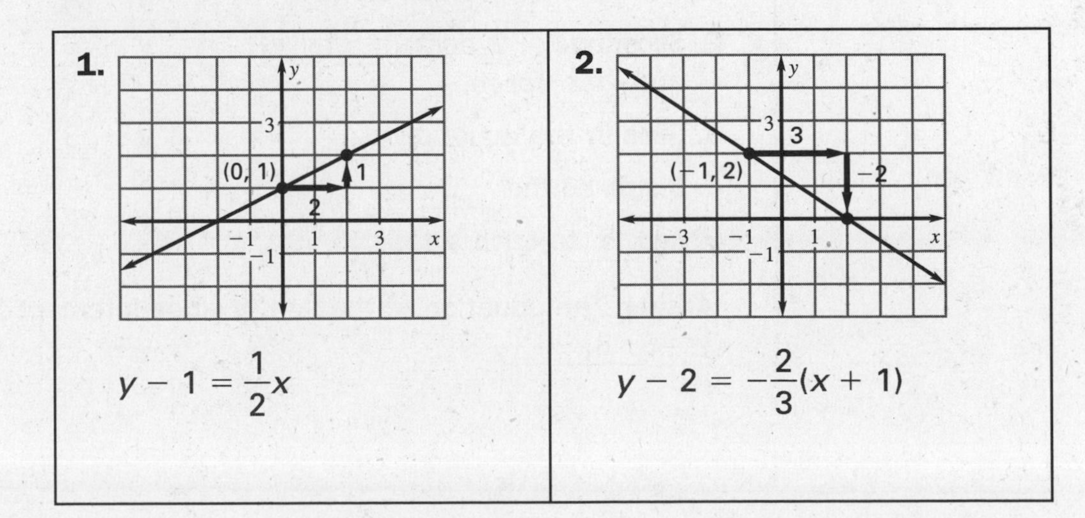

1. $y - 1 = \frac{1}{2}x$

2. $y - 2 = -\frac{2}{3}(x + 1)$

Example 2 *Write an Equation in Point-Slope Form*

Write in point-slope form the equation of the line that passes through the point (2, −4) with slope 5.

The point-slope form $y - y_1 = m(x - x_1)$ has two minus signs. Be sure to account for these signs when the point (x_1, y_1) has negative coordinates.

Solution

1. **Write** the point-slope form. $y - y_1 = m(x - x_1)$
2. **Substitute** 2 for x_1, −4 for y_1, and 5 for m. $y - (-4) = 5(x - 2)$
3. **Simplify** the equation. $y + 4 = 5(x - 2)$

Answer The equation in point-slope form of the line is $y + 4 = 5(x - 2)$.

WRITING EQUATIONS OF LINES

1. ***Use slope-intercept form***

$y = mx + b$

if you are given the slope m and the y-intercept b.

2. ***Use point-slope form***

$y - y_1 = m(x - x_1)$

if you are given the slope m and a point (x_1, y_1).

Example 3 *Use Point-Slope Form*

Write in slope-intercept form the equation of the line that passes through the point (−2, 5) with slope −3.

Solution

1. **Write** the point-slope form. $y - y_1 = m(x - x_1)$
2. **Substitute** −2 for x_1, 5 for y_1, and −3 for m. $y - 5 = -3[x - (-2)]$
3. **Simplify** the equation. $y - 5 = -3(x + 2)$
4. **Distribute** the −3. $y - 5 = -3x - 6$
5. **Add** 5 to each side. $y = -3x - 1$

Answer The equation of the line in slope-intercept form is $y = -3x - 1$.

Checkpoint **Write in point-slope form the equation of the line that passes through the given point and has the given slope. Then rewrite the equation in slope-intercept form.**

3. $(-3, -5), m = 2$	**4.** $(-1, 9), m = -2$
$y + 5 = 2(x + 3)$	$y - 9 = -2(x + 1)$
$y = 2x + 1$	$y = -2x + 7$

Example 4 ***Write an Equation of a Parallel Line***

Write in slope-intercept form the equation of the line that is parallel to the line $y = -3x + 2$ and passes through the point $(1, -5)$.

Solution

The slope of the original line is $m = \underline{-3}$. So, the slope of the parallel line is also $m = \underline{-3}$. The line passes through the point $(x_1, y_1) = (1, -5)$.

$y - y_1 = m(x - x_1)$	**Write point-slope form.**
$y - \underline{(-5)} = \underline{-3}(x - \underline{1})$	**Substitute** $\underline{-3}$ **for** m, $\underline{1}$ **for** x_1, **and** $\underline{-5}$ **for** y_1.
$y \underline{+ 5} = \underline{-3}(x - \underline{1})$	**Simplify.**
$y \underline{+ 5} = \underline{-3x + 3}$	**Use distributive property.**
$y = \underline{-3x - 2}$	**Subtract** $\underline{5}$ **from each side.**

Answer The equation of the line is $\underline{y = -3x - 2}$.

5.3 Writing Linear Equations Given Two Points

Goal Write an equation of a line given two points on the line.

Example 1 ***Use a Graph***

The line at the right models a snowboarder's descent down a mountain. Write the equation of the line in slope-intercept form.

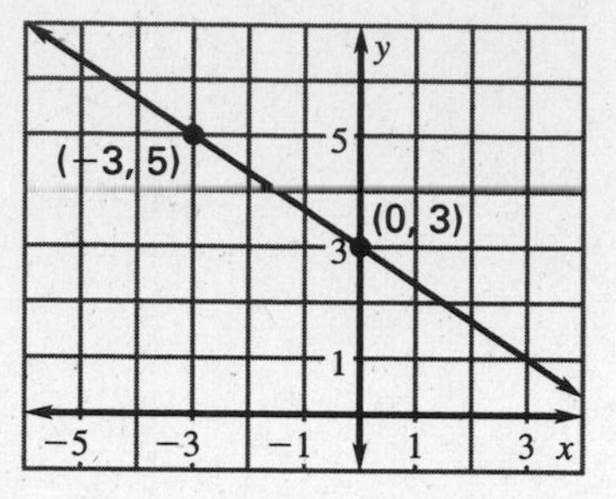

Solution

1. Find the slope.

$$m = \frac{y_2 - y_1}{x_2 - x_1} = \frac{5 - 3}{-3 - 0} = -\frac{2}{3}$$

2. Write the equation of the line. From the graph, you can see the y-intercept is $b =$ 3. Use the slope-intercept form.

$y = mx + b$ **Write slope-intercept form.**

$y = -\frac{2}{3}x + 3$ **Substitute $-\frac{2}{3}$ for m and 3 for b.**

Answer The equation of the line is $y = -\frac{2}{3}x + 3$.

Checkpoint **Write the equation of the line in slope-intercept form.**

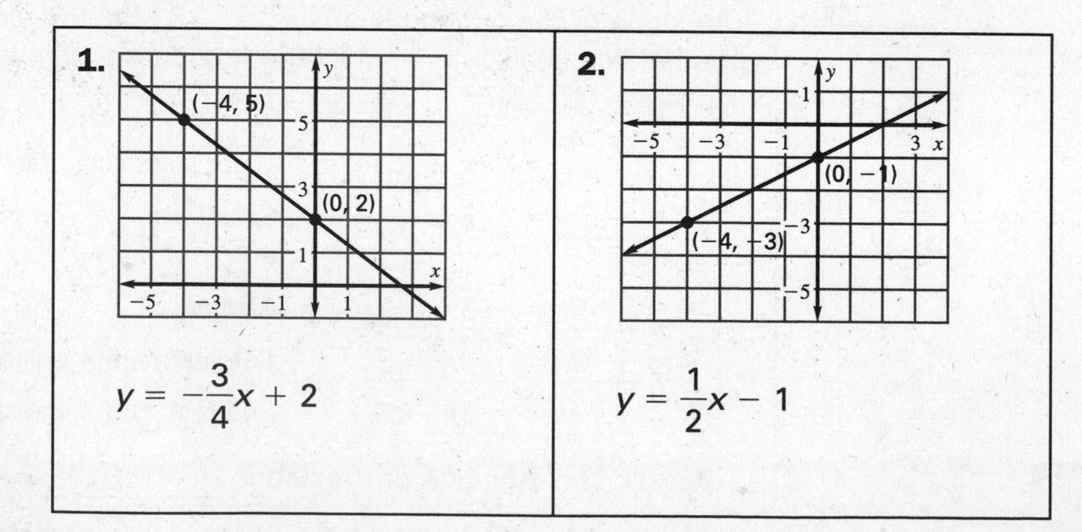

1. $y = -\frac{3}{4}x + 2$

2. $y = \frac{1}{2}x - 1$

WRITING LINEAR EQUATIONS GIVEN TWO POINTS

1. Find the slope $m = \dfrac{y_2 - y_1}{x_2 - x_1}$.

2. Write the equation of the line.

- Use the **slope-intercept** form if you know the y-intercept.
  Slope-intercept form: $y = mx + b$
- Use the **point-slope form** if you do *not* know the y-intercept.
  Point-slope form: $y - y_1 = m(x - x_1)$

Example 2 ***Write an Equation of a Line Given Two Points***

Write in slope-intercept form the equation of the line that passes through the points (2, 8) and (−5, 1).

Solution

1. Find the slope. Use $(x_1, y_1) = (2, 8)$ **and** $(x_2, y_2) = (-5, 1)$.

$m = \dfrac{y_2 - y_1}{x_2 - x_1}$ **Write formula for slope.**

$= \dfrac{1 - 8}{-5 - 2}$ **Substitute.**

$= \dfrac{-7}{-7}$ **Simplify.**

$= 1$ **Simplify.**

2. Write the equation of the line. Use point-slope form, because you do not know the ***y*****-intercept.**

$y - y_1 = m(x - x_1)$ **Write point-slope form.**

$y - 8 = 1(x - 2)$ **Substitute** 1 **for** m, 2 **for** x_1, **and** 8 **for** y_1.

$y - 8 = x - 2$ **Use distributive property.**

$y = x + 6$ **Add 8 to each side.**

Answer The equation of the line is $y = x + 6$.

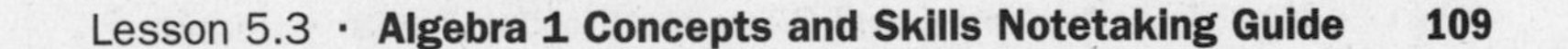

 ***Checkpoint*** **Write an equation in slope-intercept form of the line that passes through the given points.**

3. (−7, 3), (−4, 1)	**4.** (5, −6), (−2, 1)	**5.** (−2, −9), (6, −3)
$y = -\frac{2}{3}x - \frac{5}{3}$	$y = -x - 1$	$y = \frac{3}{4}x - \frac{15}{2}$

Example 3 ***Decide Which Form to Use***

Write the equation of the line in slope-intercept form that passes through the points (2, −4) and (4, 2).

Solution

Find the slope. Use $(x_1, y_1) = (2, -4)$ and $(x_2, y_2) = (4, 2)$.

$$m = \frac{y_2 - y_1}{x_2 - x_1} = \frac{2 - (-4)}{4 - 2} = \frac{6}{2} = 3$$

Since you do not know the *y*-intercept, use point-slope form.

$y - y_1 = m(x - x_1)$	**Write point-slope form.**
$y - (-4) = 3(x - 2)$	**Substitute** 3 **for** m, 2 **for** x_1, **and** −4 **for** y_1.
$y + 4 = 3(x - 2)$	**Simplify.**
$y + 4 = 3x - 6$	**Use distributive property.**
$y = 3x - 10$	**Subtract** 4 **from each side.**

Answer The equation of the line is $y = 3x - 10$.

5.4 Standard Form

Goal Write an equation of a line in standard form.

STANDARD FORM

The standard form of an equation of a line is

$Ax + By = C$, where A and B are not both zero.

Example 1 ***Write an Equation in Standard Form***

Write in standard form an equation of the line passing through (−3, 2) with a slope of −4. Use integer coefficients.

1. **Write** the point-slope form. $y - y_1 = m(x - x_1)$
2. **Substitute** −4 for m, −3 for x_1, and 2 for y_1. $y - 2 = -4[x - (-3)]$
3. **Simplify** the equation. $y - 2 = -4(x + 3)$
4. **Use** the distributive property. $y - 2 = -4x - 12$
5. **Add** 2 to each side. (Slope-intercept form) $y = -4x - 10$
6. **Add** $4x$ to each side. (Standard form) $4x + y = -10$

✔ ***Checkpoint*** **Write in standard form an equation of the line that passes through the point and has the given slope.**

1. (2, 5), $m = -3$	**2.** (−4, 6), $m = 2$
$3x + y = 11$	$-2x + y = 14$

Example 2 *Write an Equation in Standard Form*

A line intersects the axes at (2, 0) and (0, 5). Write an equation of the line in standard form. Use integer coefficients.

Solution

1. Find the slope. Use $(x_1, y_1) = (2, 0)$ and $(x_2, y_2) = (0, 5)$.

$$m = \frac{y_2 - y_1}{x_2 - x_1} = \frac{5 - 0}{0 - 2} = -\frac{5}{2}$$

2. Write an equation of the line using slope-intercept form.

$y = mx + b$	**Write slope-intercept form.**
$y = -\frac{5}{2}x + 5$	**Substitute** $-\frac{5}{2}$ **for *m* and** 5 **for *b*.**
$2y = 2\left(-\frac{5}{2}x + 5\right)$	**Multiply each side by** 2.
$2y = -5x + 10$	**Use distributive property.**
$5x + 2y = 10$	**Add** $5x$ **to each side.**

The equation is in standard form.

Checkpoint Write in standard form an equation of the line that passes through the points. Use integer coefficients.

3. (3, 7), (3, 1)	4. (−2, −4), (5, 3)	5. (−1, 8), (2, −2)
$x = 3$	$x - y = 2$	$10x + 3y = 14$

Example 3 *Equations of Horizontal and Vertical Lines*

Write the standard form of an equation of the line.

a. horizontal line

b. vertical line

Solution

a. Each point on this horizontal line has a *y*-coordinate of -2. So, the equation of the line is $y = -2$.

b. Each point on this vertical line has an *x*-coordinate of 4. So, the equation of the line is $x = 4$.

EQUATIONS OF LINES

Slope-intercept form: $y = mx + b$

Point-slope form: $y - y_1 = m(x - x_1)$

Vertical line (undefined slope): $x = a$

Horizontal line (zero slope): $y = b$

Standard form: $Ax + By = C$, where A and B are not both zero.

5.5 Modeling with Linear Equations

Goal Write and use a linear equation to solve a real-life problem.

VOCABULARY

Linear model A linear model is a linear function that is used to model a real-life situation.

Rate of change A rate of change compares two quantities that are changing.

Example 1 ***Write a Linear Model***

From 1996 through 2002, consumer Internet usage in the United States increased by about 27 hours per person per year. In 2000, consumer Internet usage was about 118 hours per person. Write a linear model for the number of hours per person y. Let $t = 0$ represent 1996.

Solution

The rate of increase is 27 hours per year, so the slope is $m =$ 27. The year 2000 is represented by $t = 4$. So, $(t_1, y_1) = ($ 4, 118 $)$ is a point on the line.

1. **Write the point-slope form.** $y - y_1 = m(t - t_1)$
2. **Substitute for m, t_1, and y_1.** $y - 118 = 27(t - 4)$
3. **Use the distributive property.** $y - 118 = 27t - 108$
4. **Add** 118 **to each side.** $y = 27t + 10$

Answer The linear model for consumer Internet usage in the United States is $y = 27t + 10$, where $t = 0$ represents 1996.

Example 2 *Use a Linear Model to Predict*

Use the linear model in Example 1 to predict the consumer Internet usage in the United States (in number of hours per person) in the year 2005.

Method 1 Use an algebraic approach. Because $t = 0$ represents the year 1996, 2005 is represented by $t = \underline{9}$.

$$y = \underline{27t + 10} = \underline{27(9) + 10} = \underline{243 + 10} = \underline{253}$$

Answer You can predict that consumer Internet usage will be about <u>253 hours</u> per person in 2005.

Method 2 Use a graphical approach. Graph the equation <u>$y = 27t + 10$</u> at the right.

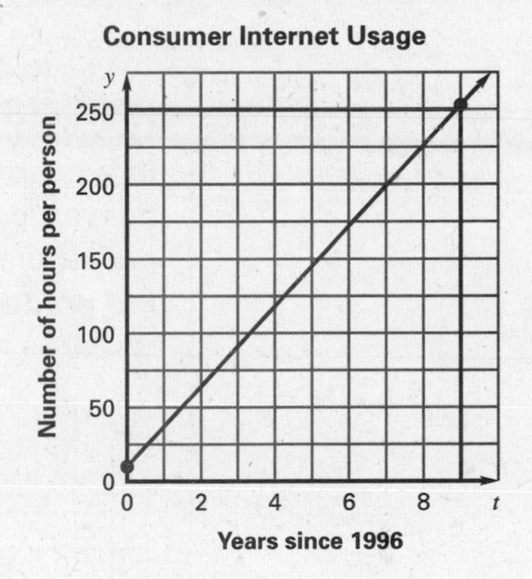

Answer From the graph, you can see that when $t = \underline{9}$, $y \approx \underline{250}$. You can predict that consumer Internet usage will be about <u>250 hours</u> per person in 2005.

On the graph, $t = 9$ represents the year 2005.

✔ *Checkpoint* **Complete the following exercises.**

1. From 1996 through 2002, video game usage in the United States increased by about 11 hours per person per year. In 1996, video game usage was about 25 hours per person. Write a linear model for the number of hours per person y. Let $t = 0$ represent 1996.

$y = 11t + 25$

2. In Exercise 1, predict video game usage in the year 2005.

About 124 hours per person

Perpendicular Lines

Goal Write equations of perpendicular lines.

VOCABULARY

Perpendicular Two lines in a plane are perpendicular if they intersect at a right, or 90°, angle.

PERPENDICULAR LINES

In a coordinate plane, two nonvertical lines are perpendicular if and only if the product of their slopes is −1.

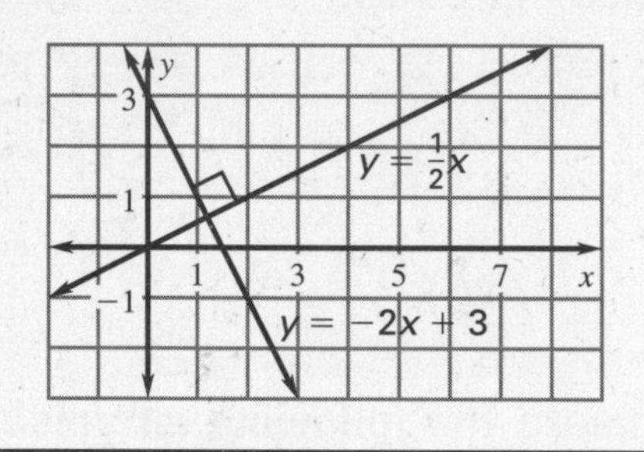

Horizontal and vertical lines are perpendicular to each other.

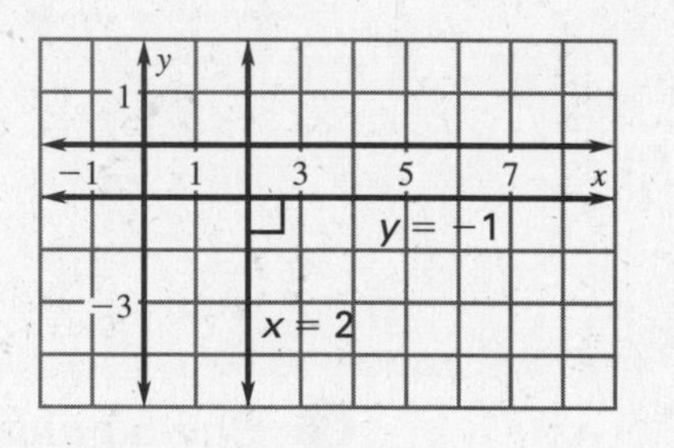

Example 1 *Identify Perpendicular Lines*

Determine whether the lines are perpendicular.

$y = -\frac{2}{3}x - 3$

$y = \frac{3}{2}x$

Solution

The lines have slopes of $\underline{-\frac{2}{3}}$ and $\underline{\frac{3}{2}}$.

Because $\left(\underline{-\frac{2}{3}}\right)\left(\underline{\frac{3}{2}}\right) = \underline{-1}$, the lines are perpendicular.

Example 2 ***Show that Lines are Perpendicular***

a. Write in slope-intercept form the equation of the line passing through (2, 3) and (6, 2).

b. Show that the line is perpendicular to the line $y = 4x + 1$.

Solution

a. ❶ **Find** the slope. Let $(x_1, y_1) = (2, 3)$ and $(x_2, y_2) = (6, 2)$.

$$m = \frac{y_2 - y_1}{x_2 - x_1} = \underline{\frac{2-3}{6-2}} = \underline{-\frac{1}{4}}$$

❷ **Write** the equation of the line using point-slope form.

$y - y_1 = m(x - x_1)$ **Write point-slope form.**

$y - \underline{3} = \underline{-\frac{1}{4}}(x - \underline{2})$ **Substitute** $\underline{-\frac{1}{4}}$ **for** m, $\underline{2}$ **for** x_1, **and** $\underline{3}$ **for** y_1.

$y - \underline{3} = \underline{-\frac{1}{4}x + \frac{1}{2}}$ **Use distributive property.**

$y = \underline{-\frac{1}{4}x + \frac{7}{2}}$ **Add** $\underline{3}$ **to each side.**

Answer The equation of the line is $\underline{y = -\frac{1}{4}x + \frac{7}{2}}$.

b. The lines have slopes of $\underline{-\frac{1}{4}}$ and $\underline{4}$. Because $\underline{\left(-\frac{1}{4}\right)(4)} = \underline{-1}$, the lines are <u>perpendicular</u>.

Checkpoint Determine whether the lines are perpendicular.

1. $y = 2x + 1, y = -2x - 1$	2. $y = 5x + 3, y = -\frac{1}{5}x - 7$
no	yes

Write in slope-intercept form the equation of the line passing through the two points. Show that the line through the points is perpendicular to the given line.

3. $(0, 3), (-2, 4); y = 2x + 1$

$y = -\frac{1}{2}x + 3$; because $\left(-\frac{1}{2}\right)(2) = -1$, the lines are perpendicular.

4. $(1, 4), (-1, -2); y = -\frac{1}{3}x - 7$

$y = 3x + 1$; because $(3)\left(-\frac{1}{3}\right) = -1$, the lines are perpendicular.

Words to Review

Give an example of the vocabulary word.

Point-slope form	**Standard form**
$y - 2 = 3(x - 4)$	$3x + 4y = 5$
Linear model	**Rate of change**
The linear model for consumer Internet usage in the United States is $y = 27t + 10$, where $t = 0$ represents 1996.	From 1996 through 2002, consumer Internet usage in the United States increased by about 27 hours per person per year.
Perpendicular	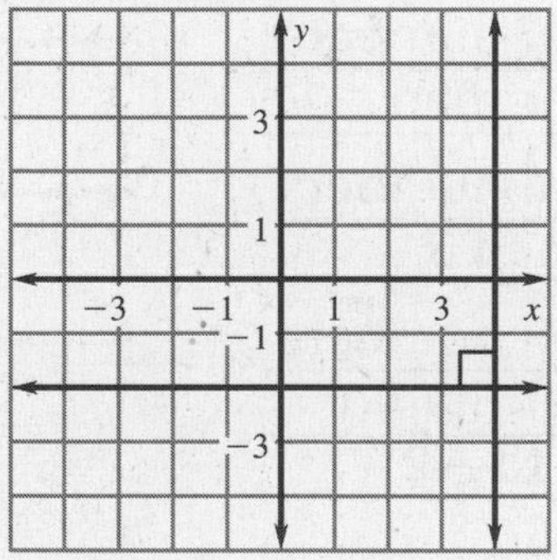
The lines $x = 4$ and $y = -2$ are perpendicular.	

Review your notes and Chapter 5 by using the Chapter Review on pages 313–316 of your textbook.

6.1 Solving Inequalities Using Addition or Subtraction

Goal Solve and graph one-step inequalities in one variable using addition or subtraction.

VOCABULARY

Graph of an inequality The graph of an inequality in one variable is the set of points on a number line that represent all solutions of the inequality.

Equivalent inequalities Equivalent inequalities are inequalities that have the same solution(s).

Example 1 *Graph an Inequality in One Variable*

INEQUALITY	VERBAL PHRASE	GRAPH
a. $b > -1$	All real numbers greater than -1	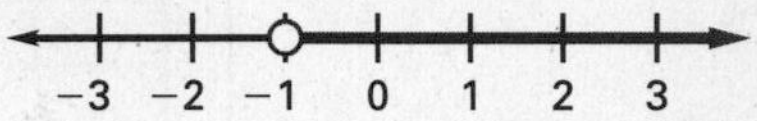
b. $v < 3$	All real numbers less than 3	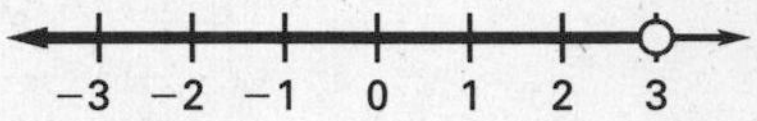
c. $f \geq -2$	All real numbers greater than or equal to -2	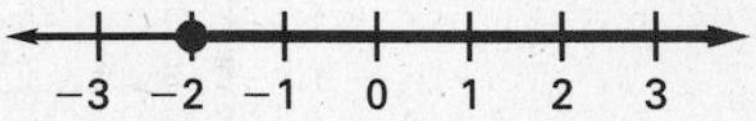

The properties are stated for > and < inequalities. They are also true for ≥ and ≤ inequalities.

PROPERTIES OF INEQUALITY

Addition Property of Inequality

For all real numbers a, b, and c: If $a > b$, then $a + c > b + c$.
If $a < b$, then $a + c < b + c$.

Subtraction Property of Inequality

For all real numbers a, b, and c: If $a > b$, then $a - c > b - c$.
If $a < b$, then $a - c < b - c$.

Example 2 *Use Subtraction to Solve an Inequality*

Solve $x + 11 \le 7$. Then graph the solution.

$x + 11 \le 7$	**Write original inequality.**
$x + 11 - \underline{11} \le 7 - \underline{11}$	**Subtract $\underline{11}$ from each side.**
$x \le \underline{-4}$	**Simplify.**

Answer The solution is all real numbers $\underline{\text{less than or equal to } -4}$.

−10 −8 −6 −4 −2 0 2 4 6 8 10 12 14 16 18 20 22

Example 3 *Use Addition to Solve an Inequality*

Solve $-8 > n - 19$. Then graph the solution.

$-8 > n - 19$	**Write original inequality.**
$-8 + \underline{19} > n - 19 + \underline{19}$	**Add $\underline{19}$ to each side.**
$\underline{11} > n$	**Simplify.**

Answer The solution is all real numbers $\underline{\text{less than } 11}$.

0 1 2 3 4 5 6 7 8 9 10 11 12 13

Checkpoint **Solve the inequality. Then graph the solution.**

1. $y + 5 \le 8$ −1 0 1 2 3 4 5 $y \le 3$	**2.** $x - 4 > -13$ −12 −11 −10 −9 −8 −7 −6 −5 $x > -9$

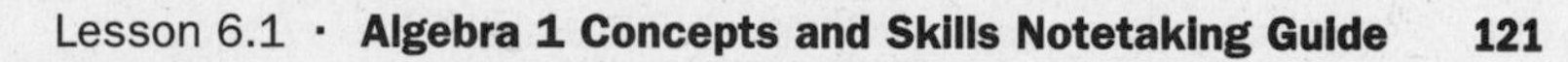

6.2 Solving Inequalities Using Multiplication or Division

Goal Solve and graph one-step inequalities in one variable using multiplication or division.

> The properties are stated for > and < inequalities. They are also true for ≥ and ≤ inequalities.

PROPERTIES OF INEQUALITY

Multiplication Property of Inequality ($c > 0$)

For all real numbers a, b, and for $c > 0$: If $a > b$, then $ac > bc$.
If $a < b$, then $ac < bc$.

Division Property of Inequality ($c > 0$)

For all real numbers a, b, and for $c > 0$: If $a > b$, then $\frac{a}{c} > \frac{b}{c}$.
If $a < b$, then $\frac{a}{c} < \frac{b}{c}$.

Example 1 ***Multiply by a Positive Number***

Solve $\frac{b}{5} \geq 12$. Then graph the solution.

Solution

$\frac{b}{5} \geq 12$ **Write original inequality.**

$5 \cdot \frac{b}{5} \geq 5 \cdot 12$ **Multiply each side by 5.**

$b \geq 60$ **Simplify.**

Answer The solution is all real numbers greater than or equal to 60.

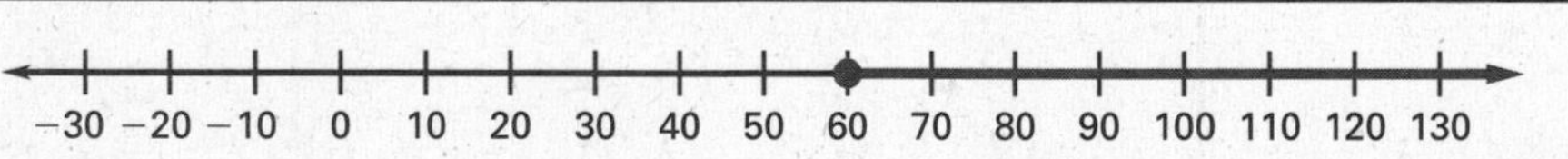

Example 2 *Divide by a Positive Number*

Solve $-54 > 9n$. Then graph the solution.

Solution

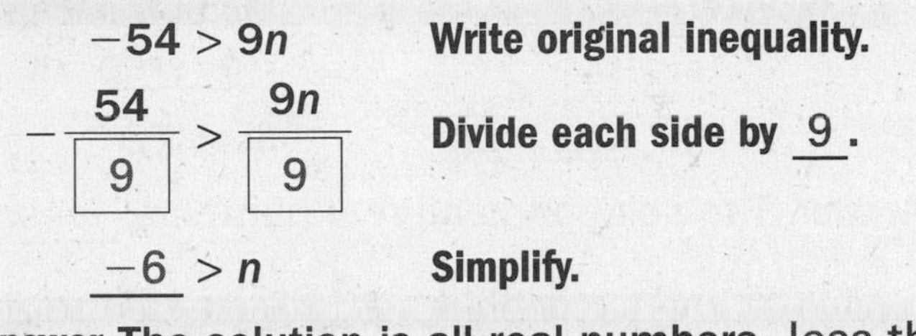

$-54 > 9n$ **Write original inequality.**

$-\frac{54}{\boxed{9}} > \frac{9n}{\boxed{9}}$ **Divide each side by** $\underline{9}$.

$\underline{-6} > n$ **Simplify.**

Answer The solution is all real numbers $\underline{\text{less than } -6}$.

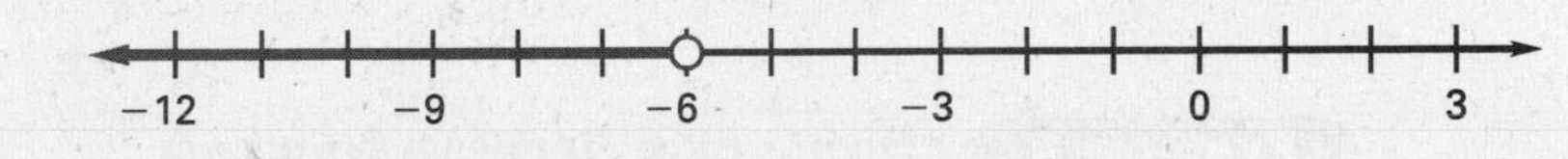

Checkpoint Solve the inequality. Then graph the solution.

1. $\frac{x}{5} < -3$	**2.** $8y \geq 48$
−18 −17 −16 −15 −14 −13 −12	2 3 4 5 6 7 8 9
$x < -15$	$y \geq 6$

Remember when multiplying or dividing by a negative number you must reverse the inequality symbol.

PROPERTIES OF INEQUALITY

Multiplication Property of Inequality ($c < 0$)

For all real numbers a, b, and for $c < 0$: If $a > b$, then $\underline{ac < bc}$.

If $a < b$, then $\underline{ac > bc}$.

Division Property of Inequality ($c < 0$)

For all real numbers a, b, and for $c < 0$: If $a > b$, then $\underline{\frac{a}{c} < \frac{b}{c}}$.

If $a < b$, then $\underline{\frac{a}{c} > \frac{b}{c}}$.

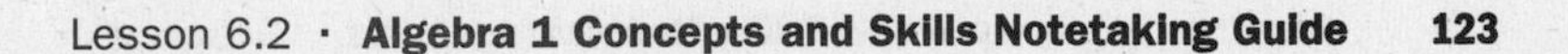

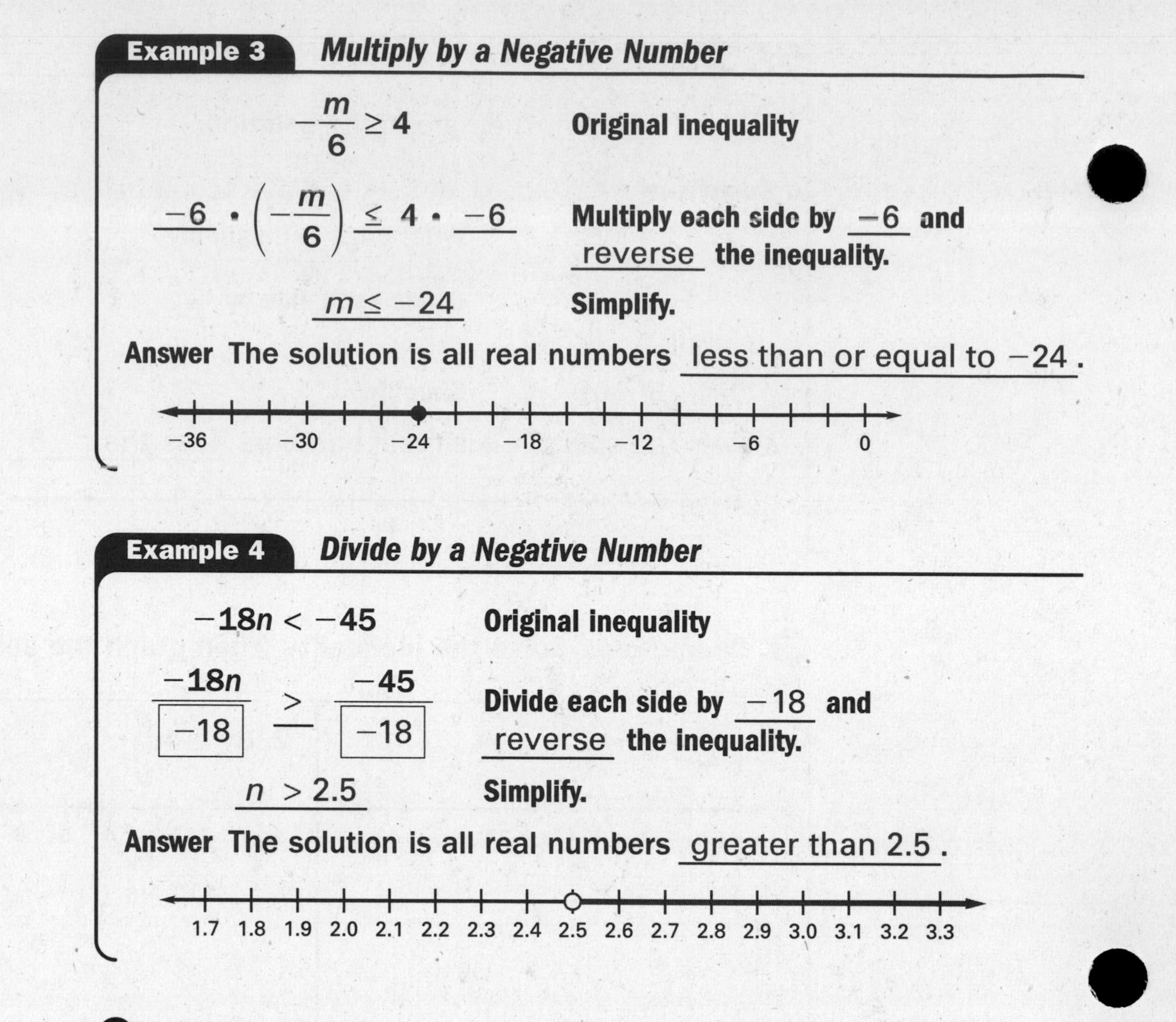

Example 3 ***Multiply by a Negative Number***

$-\frac{m}{6} \geq 4$ **Original inequality**

$\underline{-6} \cdot \left(-\frac{m}{6}\right) \underline{\leq} 4 \cdot \underline{-6}$ **Multiply each side by** $\underline{-6}$ **and** $\underline{\text{reverse}}$ **the inequality.**

$\underline{m \leq -24}$ **Simplify.**

Answer The solution is all real numbers $\underline{\text{less than or equal to } -24}$.

Example 4 ***Divide by a Negative Number***

$-18n < -45$ **Original inequality**

$\frac{-18n}{\boxed{-18}} \underline{>} \frac{-45}{\boxed{-18}}$ **Divide each side by** $\underline{-18}$ **and** $\underline{\text{reverse}}$ **the inequality.**

$\underline{n > 2.5}$ **Simplify.**

Answer The solution is all real numbers $\underline{\text{greater than } 2.5}$.

✔ ***Checkpoint*** **Solve the inequality. Then graph the solution.**

3. $-10y \geq 50$	**4.** $-6 > -\frac{d}{3}$
−8 −7 −6 −5 −4 −3 −2 −1	−3 0 3 6 9 12 15 18 21 24
$y \leq -5$	$d > 18$

6.3 Solving Multi-Step Inequalities

Goal Solve multi-step inequalities in one variable.

Example 1 ***Solve a Multi-Step Inequality***

Solve $3x - 8 > 10$.

$3x - 8 > 10$	**Write original inequality.**
$3x > \underline{18}$	**Add $\underline{8}$ to each side.**
$x > \underline{6}$	**Divide each side by $\underline{3}$.**

Answer The solution is all real numbers greater than 6.

Example 2 ***Solve a Multi-Step Inequality***

Solve $12 - 5y < -8$.

$12 - 5y < -8$	**Write original inequality.**
$-5y < \underline{-20}$	**Subtract $\underline{12}$ from each side.**
$y > \underline{4}$	**Divide each side by $\underline{-5}$ and reverse the inequality.**

Answer The solution is all real numbers greater than 4.

Example 3 ***Use the Distributive Property***

Solve $4(x - 3) \geq 32$.

$4(x - 3) \geq 32$	**Write original inequality.**
$\underline{4}x - \underline{12} \geq 32$	**Use distributive property.**
$\underline{4}x \geq \underline{44}$	**Add $\underline{12}$ to each side.**
$x \geq \underline{11}$	**Divide each side by $\underline{4}$.**

Answer The solution is all real numbers greater than or equal to 11.

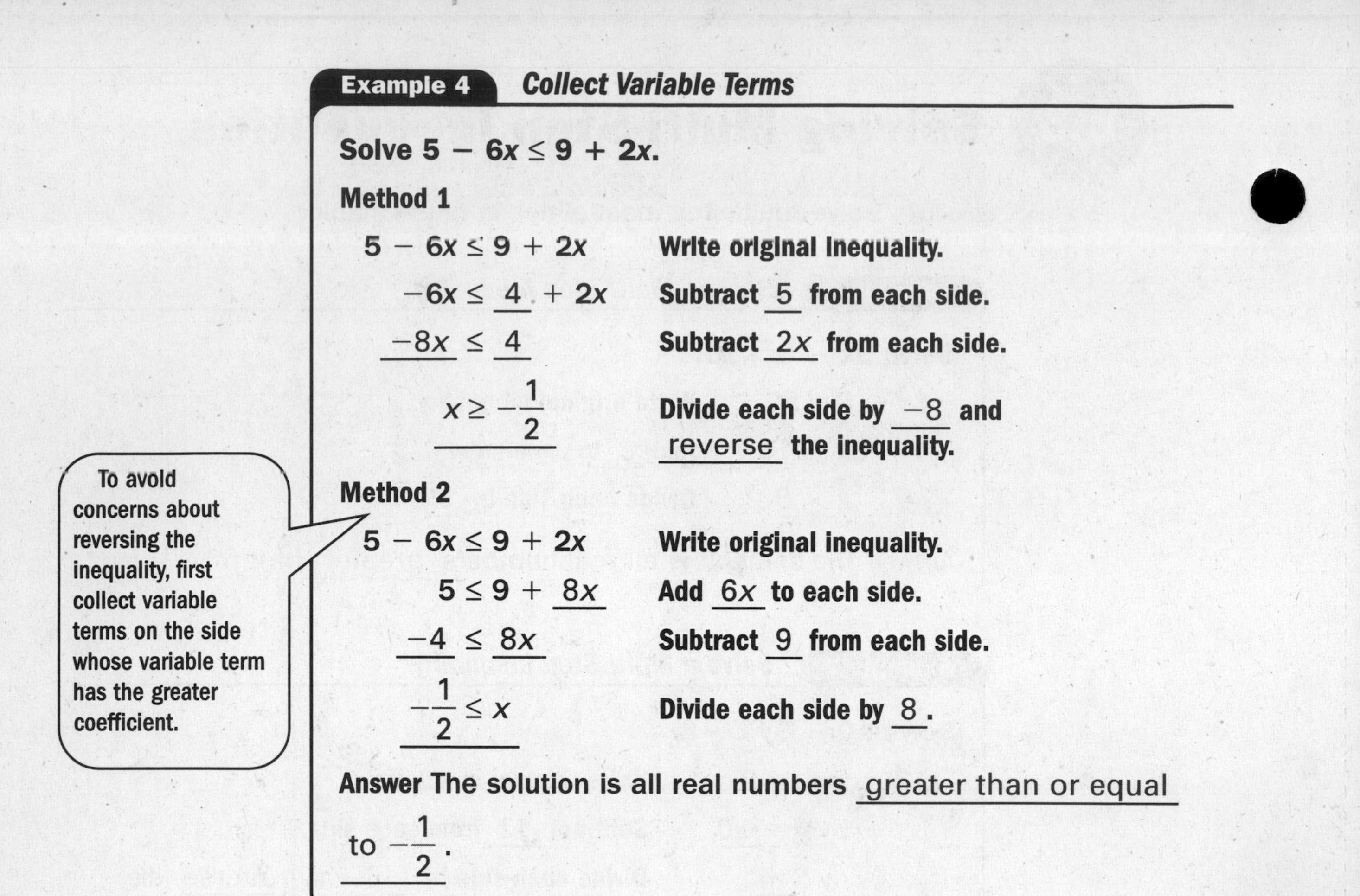

Example 4 ***Collect Variable Terms***

Solve $5 - 6x \leq 9 + 2x$.

Method 1

$5 - 6x \leq 9 + 2x$	**Write original inequality.**
$-6x \leq \underline{4} + 2x$	**Subtract** $\underline{5}$ **from each side.**
$\underline{-8x} \leq \underline{4}$	**Subtract** $\underline{2x}$ **from each side.**
$\underline{x \geq -\frac{1}{2}}$	**Divide each side by** $\underline{-8}$ **and** <u>reverse</u> **the inequality.**

> To avoid concerns about reversing the inequality, first collect variable terms on the side whose variable term has the greater coefficient.

Method 2

$5 - 6x \leq 9 + 2x$	**Write original inequality.**
$5 \leq 9 + \underline{8x}$	**Add** $\underline{6x}$ **to each side.**
$\underline{-4} \leq \underline{8x}$	**Subtract** $\underline{9}$ **from each side.**
$\underline{-\frac{1}{2} \leq x}$	**Divide each side by** $\underline{8}$**.**

Answer The solution is all real numbers <u>greater than or equal to $-\frac{1}{2}$</u>.

✔ *Checkpoint* **Solve the inequality.**

1. $9 \geq -7 - 4k$ $-4 \leq k$	**2.** $7t + 4 < -10$ $t < -2$
3. $-3(x + 4) \leq 9$ $x \geq -7$	**4.** $14 - y > 4y - 11$ $y < 5$

6.4 Solving Compound Inequalities Involving "*And*"

Goal Solve and graph compound inequalities involving *and*.

VOCABULARY

Compound inequality A compound inequality is two inequalities connected by the word *and* or the word *or*.

Example 1 ***Write Compound Inequalities with* And**

Write a compound inequality that represents the set of all real numbers greater than −3 and less than or equal to 2. Then graph the inequality.

A number is a solution of a compound inequality with *and* if the number is a solution of *both* inequalities.

Solution

The set can be represented by two inequalities.

$-3 < x$ *and* $x \leq 2$

The two inequalities can than be combined in a single inequality.

$-3 \underline{<} x \underline{\leq} 2$

The compound inequality may be read in these two ways:

- x is greater than −3 and x is less than or equal to 2.
- x is greater than −3 and less than or equal to 2.

The graph of this compound inequality is shown below.

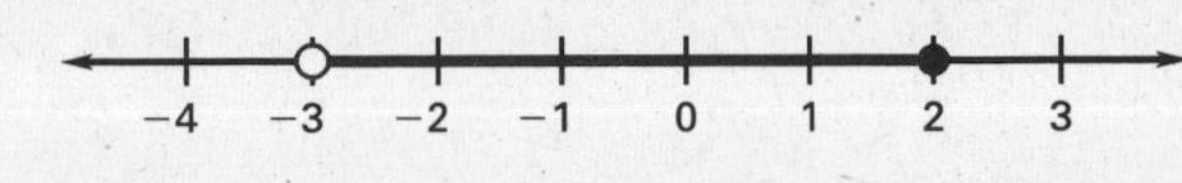

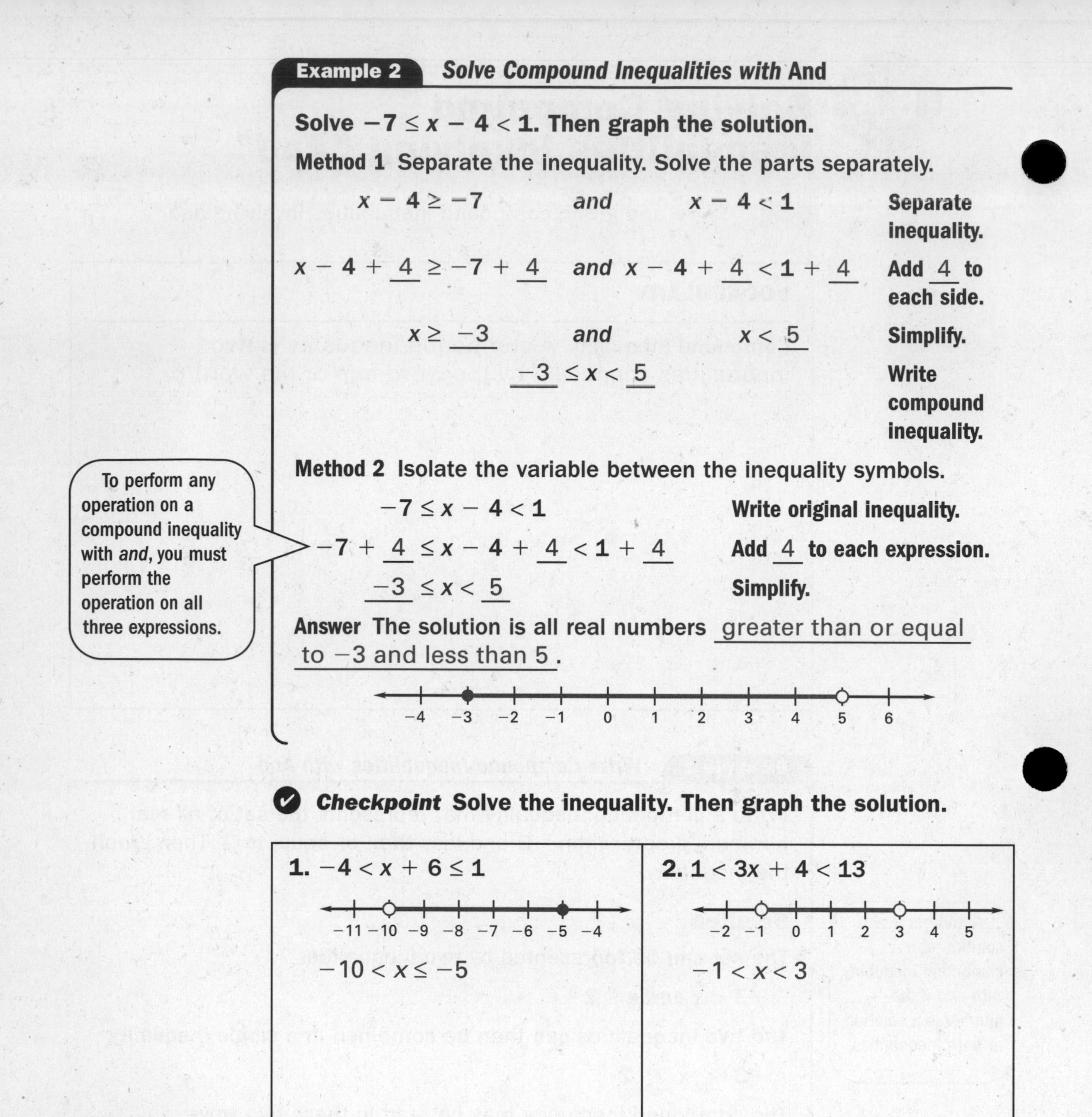

Example 2 — Solve Compound Inequalities with And

Solve $-7 \le x - 4 < 1$. Then graph the solution.

Method 1 Separate the inequality. Solve the parts separately.

$x - 4 \ge -7$	*and*	$x - 4 < 1$	**Separate inequality.**
$x - 4 + \underline{4} \ge -7 + \underline{4}$	*and*	$x - 4 + \underline{4} < 1 + \underline{4}$	**Add $\underline{4}$ to each side.**
$x \ge \underline{-3}$	*and*	$x < \underline{5}$	**Simplify.**
	$\underline{-3} \le x < \underline{5}$		**Write compound inequality.**

Method 2 Isolate the variable between the inequality symbols.

$-7 \le x - 4 < 1$	**Write original inequality.**
$-7 + \underline{4} \le x - 4 + \underline{4} < 1 + \underline{4}$	**Add $\underline{4}$ to each expression.**
$\underline{-3} \le x < \underline{5}$	**Simplify.**

> To perform any operation on a compound inequality with *and*, you must perform the operation on all three expressions.

Answer The solution is all real numbers greater than or equal to -3 and less than 5.

✔ ***Checkpoint*** **Solve the inequality. Then graph the solution.**

1. $-4 < x + 6 \le 1$	**2.** $1 < 3x + 4 < 13$
−11 −10 −9 −8 −7 −6 −5 −4	−2 −1 0 1 2 3 4 5
$-10 < x \le -5$	$-1 < x < 3$

Example 3 *Reverse Both Inequalities*

Solve $-7 \le -2x + 1 < -3$. Then graph the solution.

Isolate the variable between the two inequality symbols.

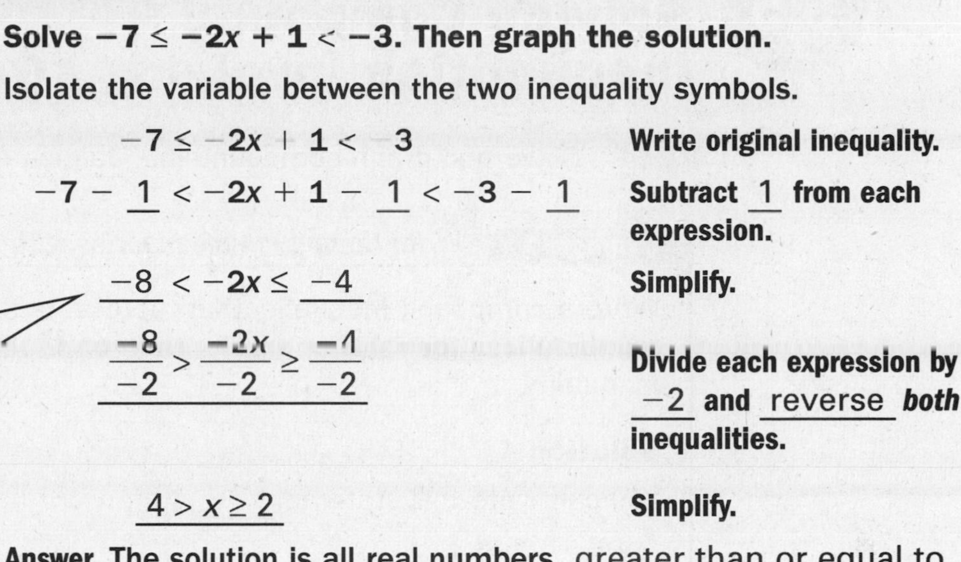

$-7 \le -2x + 1 < -3$	**Write original inequality.**
$-7 - \underline{1} < -2x + 1 - \underline{1} < -3 - \underline{1}$	**Subtract** $\underline{1}$ **from each expression.**
$\underline{-8} < -2x \le \underline{-4}$	**Simplify.**
$\frac{-8}{-2} > \frac{-2x}{-2} \ge \frac{-4}{-2}$	**Divide each expression by** $\underline{-2}$ **and** $\underline{\text{reverse}}$ ***both*** **inequalities.**
$4 > x \ge 2$	**Simplify.**

When you multiply or divide each expression of a compound inequality by a negative number, remember to reverse *both* inequalities.

Answer The solution is all real numbers greater than or equal to 2 and less than 4.

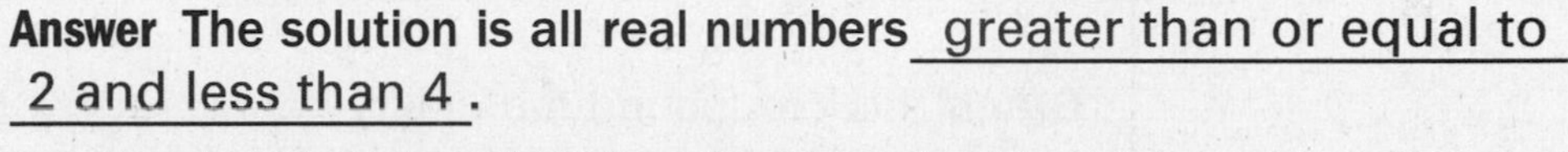

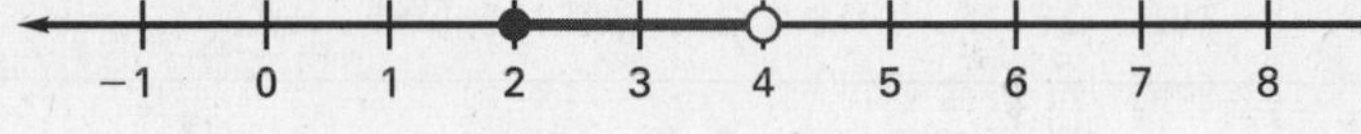

Checkpoint Solve the inequality. Then graph the solution.

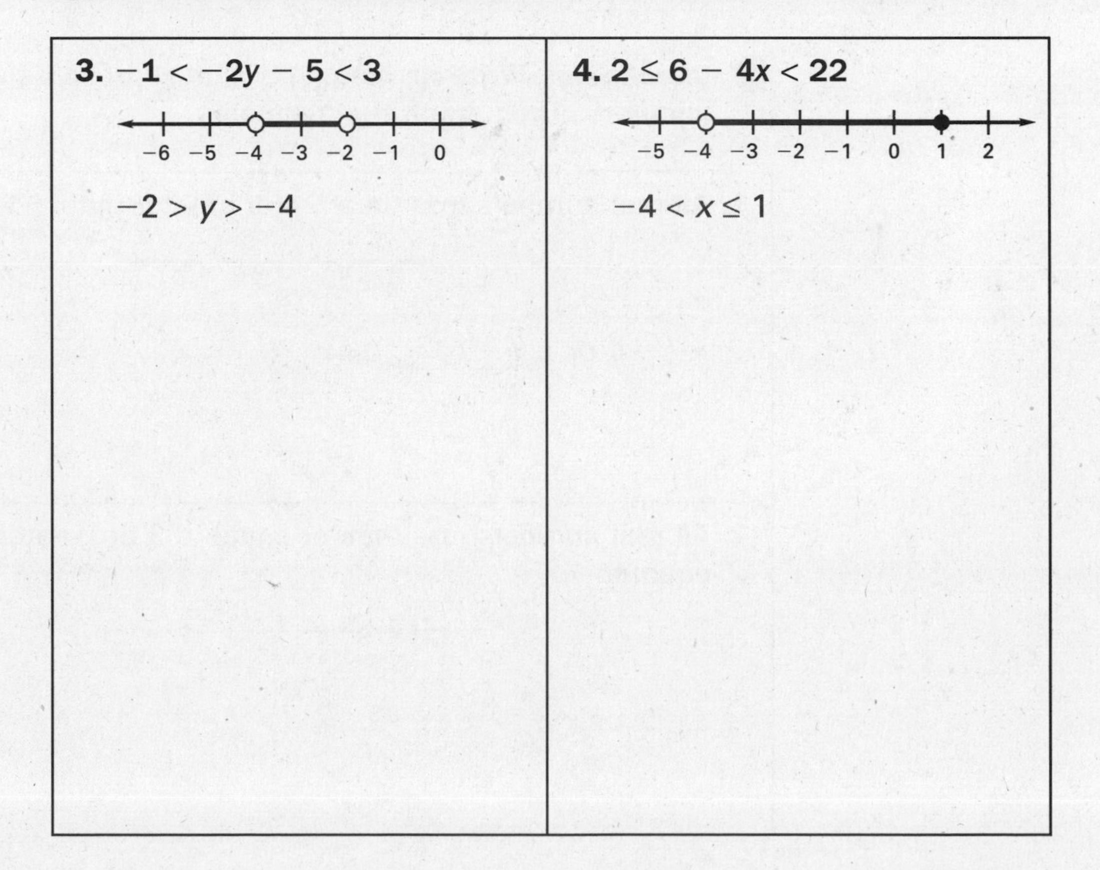

3. $-1 < -2y - 5 < 3$	**4.** $2 \le 6 - 4x < 22$
$-2 > y > -4$	$-4 < x \le 1$

6.5 Solving Compound Inequalities Involving "*Or*"

Goal Solve and graph compound inequalities involving *or*.

Example 1 ***Write Compound Inequalities with* Or**

Write a compound inequality that represents the set of all real numbers less than −3 or greater than or equal to 3. Then graph the inequality.

Solution

You can write this statement using the word *or*.

x <u>$<$</u> −3 or x <u>$\geq$</u> 3

Graph this compound inequality below. Notice that the graph has two parts. One part lies to the <u>left of −3</u>. The other part lies to the <u>right of 3</u>.

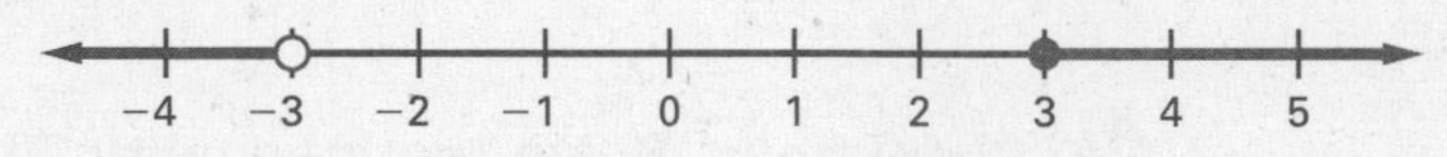

✓ ***Checkpoint*** **Write an inequality that represents the set of numbers. Then graph the inequality.**

1. All real numbers less than −4 or greater than −1.

$x < -4$ *or* $x > -1$

2. All real numbers less than or equal to 3 or greater than or equal to 7.

$x \leq 3$ *or* $x \geq 7$

Example 2 *Solve a Compound Inequality with Or*

Solve the compound inequality $x + 7 < 3$ *or* $4x \geq 8$. Then graph the solution.

A number is a solution of a compound inequality with *or* if the number is a solution of *either* inequality.

Solution

A solution of this compound inequality is a solution of either of its parts. You can solve each part separately using the methods of Lessons 6.1 and 6.2.

$x + 7 < 3$	*or*	$4x \geq 8$	**Write original inequality.**
$x + 7 - \underline{7} < 3 - \underline{7}$	*or*	$\frac{4x}{\boxed{4}} \geq \frac{8}{\boxed{4}}$	**Isolate *x*.**
$x < \underline{-4}$	*or*	$x \geq \underline{2}$	**Simplify.**

Answer The solution is all real numbers less than −4 or greater than or equal to 2. Graph the solution below.

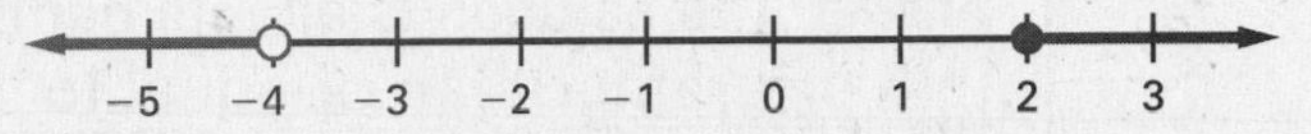

Checkpoint Solve the inequality. Then graph the solution.

3. $x + 4 < -1$ *or* $3x > 45$

−10 −5 0 5 10 15 20

$x < -5$ *or* $x > 15$

4. $\frac{x}{2} \leq -4$ *or* $x + 7 > 3$

−10 −8 −6 −4 −2 0 2

$x \leq -8$ *or* $x > -4$

Example 3 *Solve a Multi-Step Compound Inequality*

Solve the compound inequality $2 - 5x \le -3$ or $-4x - 3 \ge 17$. Then graph the solution.

Solution

Solve each of the parts using the methods of Lesson 6.3.

$2 - 5x \le -3$	*or*	$-4x - 3 \ge 17$	**Original inequality**
$2 - 5x - \underline{2} \le -3 - \underline{2}$	*or*	$-4x - 3 + \underline{3} \ge 17 + \underline{3}$	**Isolate *x*.**
$-5x \le \underline{-5}$	*or*	$-4x \ge \underline{20}$	**Simplify.**
$\frac{-5x}{-5} \ge \frac{-5}{-5}$	*or*	$\frac{-4x}{-4} \le \frac{20}{-4}$	**Solve for *x*.**
$\underline{x \ge 1}$	*or*	$\underline{x \le -5}$	**Simplify.**

Answer The solution is all real numbers less than or equal to -5 or greater than or equal to 1. Graph the solution below.

−6 −5 −4 −3 −2 −1 0 1 2

Checkpoint Solve the inequality. Then graph the solution.

5. $3x - 6 \le -12$ *or* $2x + 2 > 10$

−3 −2 −1 0 1 2 3 4 5

$x \le -2$ *or* $x > 4$

6. $-9x - 5 > 13$ *or* $2x - 1 \ge 3$

−3 −2 −1 0 1 2 3 4 5

$x < -2$ *or* $x \ge 2$

6.6 Solving Absolute-Value Equations

Goal Solve absolute-value equations in one variable.

VOCABULARY

Absolute-value equation An absolute-value equation is an equation of the form $|ax + b| = c$.

SOLVING AN ABSOLUTE-VALUE EQUATION

For $c \geq 0$, x is a solution of $|ax + b| = c$ if x is a solution of:

$ax + b = c$ *or* $ax + b = -c$

For $c < 0$, the absolute-value equation $|ax + b| = c$ has no solution, since absolute value always indicates a number that is not negative.

Example 1 ***Solve an Absolute-Value Equation***

Solve the equation.

a. $|x| = 4$ **b.** $|x| = -11$

Solution

a. There are two values of x that have an absolute value of 4.

$|x| = 4$

$x = 4$ *or* $x = -4$

Answer The equation has two solutions: 4 and -4.

b. The absolute value of a number is never negative.

Answer The equation $|x| = -11$ has no solution.

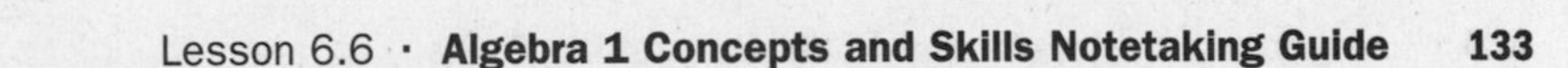

Example 2 *Solve an Absolute-Value Equation*

Solve $|x + 4| = 7$.

Solution

Because $|x + 4| = 7$, the expression $x + 4$ is equal to 7 *or* −7.

$x + 4$ is positive *or* **$x + 4$ is negative**

$$x + 4 = 7 \quad \text{or} \quad x + 4 = -7$$

$$x + 4 - 4 = 7 - 4 \quad \quad x + 4 - 4 = -7 - 4$$

$$x = 3 \quad \text{or} \quad x = -11$$

Answer The equation has two solutions: 3 and −11.

Example 3 *Solve an Absolute-Value Equation*

Solve $|4x - 6| - 7 = -5$.

Solution

First isolate the absolute-value expression on one side of the equation.

$$|4x - 6| - 7 = -5$$

$$|4x - 6| - 7 + 7 = -5 + 7$$

$$|4x - 6| = 2$$

Because $|4x - 6| = 2$, the expression $4x - 6$ is equal to 2 or −2.

$4x - 6$ is positive *or* **$4x - 6$ is negative**

$$4x - 6 = 2 \quad \quad 4x - 6 = -2$$

$$4x - 6 + 6 = 2 + 6 \quad \quad 4x - 6 + 6 = -2 + 6$$

$$4x = 8 \quad \quad 4x = 4$$

$$\frac{4x}{4} = \frac{8}{4} \quad \quad \frac{4x}{4} = \frac{4}{4}$$

$$x = 2 \quad \text{or} \quad x = 1$$

Answer The equation has two solutions: 2 and 1.

Example 4 *Write an Absolute-Value Equation*

Write an absolute-value equation that has 4 and 10 as its solutions.

Solution

Graph the numbers and locate the midpoint of the graphs.

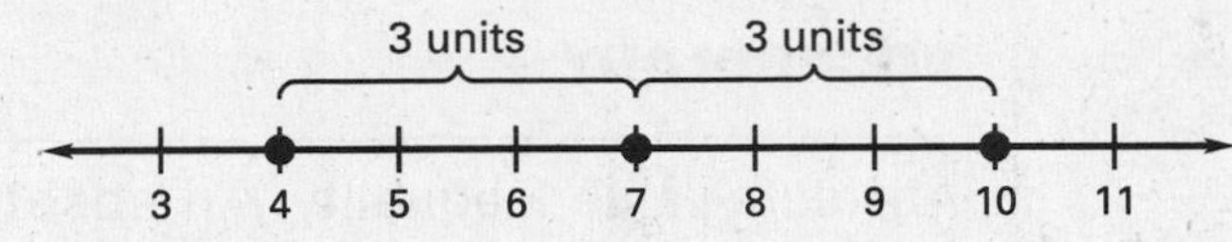

The graph of each solution is 3 units from the midpoint, 7. You can use the midpoint and the distance to write an absolute-value equation.

Midpoint ↓ **Distance** ↓

$$|x - 7| = 3$$

Answer The equation is $|x - 7| = 3$.

Checkpoint Solve the absolute-value equation.

1. $|x| = 7$ $-7, 7$	**2.** $|x| = -9$ no solution
3. $|x + 9| = 14$ $5, -23$	**4.** $|x - 8| = 4$ $12, 4$

5. Write an absolute-value equation that has 2 and 8 as it solutions.

$|x - 5| = 3$

6.7 Solving Absolute-Value Inequalities

Goal Solve absolute-value inequalities in one variable.

VOCABULARY

Absolute-value inequality An absolute-value inequality is an inequality that has one of these forms: $|ax + b| < c$, $|ax + b| \leq c$, $|ax + b| > c$, $|ax + b| \geq c$.

SOLVING ABSOLUTE-VALUE INEQUALITIES

Each absolute-value inequality is rewritten as two inequalities joined by *and* or *or*.

- $|ax + b| < c$ **means** $ax + b < c$ ***and*** $ax + b > -c$.
- $|ax + b| > c$ **means** $ax + b > c$ ***or*** $ax + b < -c$.

Similar rules apply for ≤ and ≥.

Example 1 *Solve an Absolute-Value Inequality*

Solve $|x - 3| > 5$. Then graph the solution.

The solution consists of all numbers x whose distance from 3 is greater than 5. The inequality involves > so the related inequalities are connected by *or*.

$|x - 3| > 5$			**Write original inequality.**
$x - 3 > 5$	*or*	$x - 3 < -5$	**Write related inequalities.**
$x - 3 + \underline{3} > 5 + \underline{3}$	*or*	$x - 3 + \underline{3} < -5 + \underline{3}$	**Add $\underline{3}$ to each side.**
$x > \underline{8}$	*or*	$x < \underline{-2}$	**Simplify.**

Answer **The solution is all real numbers** greater than 8 or less than −2**. This can be written as** $x < -2$ or $x > 8$**. Graph the solution below.**

−3 −2 −1 0 1 2 3 4 5 6 7 8 9

Example 2 *Solve a Multi-Step Inequality*

Solve $|x - 2| - 6 < 1$. Then graph the solution.

Solution

First isolate the absolute-value expression on one side of the inequality.

$\lvert x - 2\rvert - 6 < 1$	**Write original inequality.**
$\lvert x - 2\rvert - 6 + \underline{6} < 1 + \underline{6}$	**Add $\underline{6}$ to each side.**
$\lvert x - 2\rvert < \underline{7}$	**Simplify.**

The inequality involves < so the related inequalities are connected by *and*.

	$\lvert x - 2\rvert < \underline{7}$		**Write simplified inequality.**
$x - 2 < \underline{7}$	*and*	$x - 2 > \underline{-7}$	**Write related inequalities.**
$x - 2 + \underline{2} < \underline{7} + \underline{2}$	*and*	$x - 2 + \underline{2} > \underline{-7} + \underline{2}$	**Add $\underline{2}$ to each side.**
$x < \underline{9}$	*and*	$x > \underline{-5}$	**Simplify.**

Answer The solution is all real numbers greater than −5 and less than 9. This can be written as $-5 < x < 9$. Graph the solution below.

−6 −5 −4 −3 −2 −1 0 1 2 3 4 5 6 7 8 9 10

Checkpoint Solve the absolute-value inequality.

1. $\lvert x - 7\rvert > 3$ $x > 10$ *or* $x < 4$	**2.** $\lvert x + 5\rvert - 4 \le 9$ $-18 \le x \le 8$

6.8 Graphing Linear Inequalities in Two Variables

Goal Solve linear inequalities in two variables.

VOCABULARY

Linear inequality in two variables A linear inequality in x and y is an inequality that can be written as follows: $ax + by < c$, $ax + by \leq c$, $ax + by > c$, or $ax + by \geq c$.

Example 1 ***Check Solutions of a Linear Inequality***

Check whether the ordered pair is a solution of $8x - 4y \geq 3$.

a. (0, 0)

b. (1, −1)

Solution

(x, y)	$8x - 4y \geq 3$	Conclusion
a. (0, 0)	$8(\underline{0}) - 4(\underline{0}) = \underline{0}\ \underline{\ngeq}\ 3$	(0, 0) <u>is not</u> a solution.
b. (1, −1)	$8(\underline{1}) - 4(\underline{-1}) = \underline{12}\ \underline{\geq}\ 3$	(1, −1) <u>is</u> a solution.

GRAPHING A LINEAR INEQUALITY

Step 1 **Graph** the corresponding equation. Use a <u>dashed</u> line for > or <. Use a <u>solid</u> line for ≥ or ≤.

Step 2 **Test** the coordinates of a point in one of the <u>half-planes</u>.

Step 3 **Shade** the <u>half-plane</u> containing the point if it is a solution of the inequality. If it is not a solution, <u>shade</u> the other <u>half-plane</u>.

Example 2 *Vertical Lines*

Graph the inequality $x > 3$.

Solution

1. **Graph** the corresponding equation $x = 3$. The graph of $x = 3$ is a vertical line . The inequality is >, so use a dashed line.

2. **Test** a point. The origin (0, 0) is not a solution and it lies to the left of the line. So, the graph of $x > 3$ is all points to the right of the line $x = 3$.

3. Shade the region to the right of the line.

Answer The graph of $x > 3$ is the region to the right of the graph of $x = 3$.

Example 3 *Horizontal Lines*

Graph the inequality $y \leq -4$.

Solution

1. **Graph** the corresponding equation $y = -4$. The graph of $y = -4$ is a horizontal line. The inequality is ≤, so use a solid line.

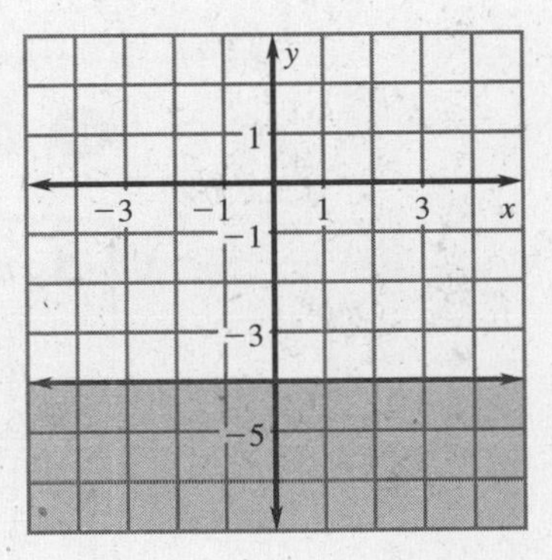

2. **Test** a point. The origin (0, 0) is not a solution and it lies above the line. So, the graph of $y \leq -4$ is all points on or below the line $y = -4$.

3. **Shade** the region below the line.

Answer The graph of $y \leq -4$ is the graph of $y = -4$ and the region below the graph of $y = -4$.

Example 4 *Use Slope-Intercept Form*

Graph the inequality $2x + 3y < 6$ using the slope-intercept form of the corresponding equation.

Solution

Write the corresponding equation in slope-intercept form.

$2x + 3y = 6$	**Write corresponding equation.**
$3y = -2x + 6$	**Subtract $2x$ from each side.**
$y = -\frac{2}{3}x + 2$	**Divide each side by 3.**

The graph of the line has a slope of $-\frac{2}{3}$ and a y-intercept of 2. The inequality is $<$, so use a dashed line.

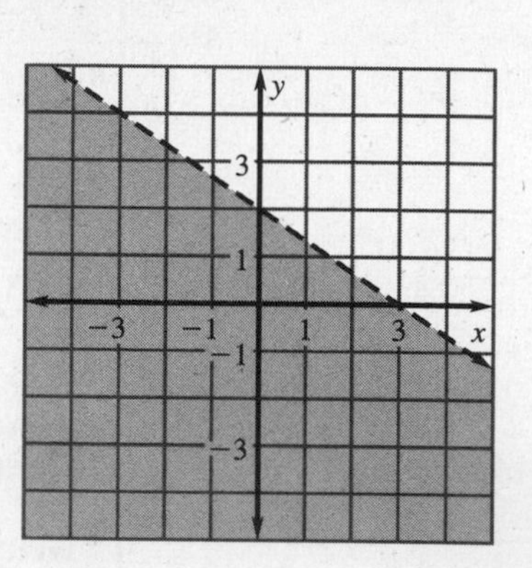

Test the origin: $2(0) + 3(0) = 0$ is a solution and it lies below the line. Since (0, 0) lies below the line, shade the region below the line.

Answer The graph of $2x + 3y < 6$ is all points below the line.

Checkpoint Graph the inequality.

1. $x + y \geq -2$

2. $5x - y < 3$

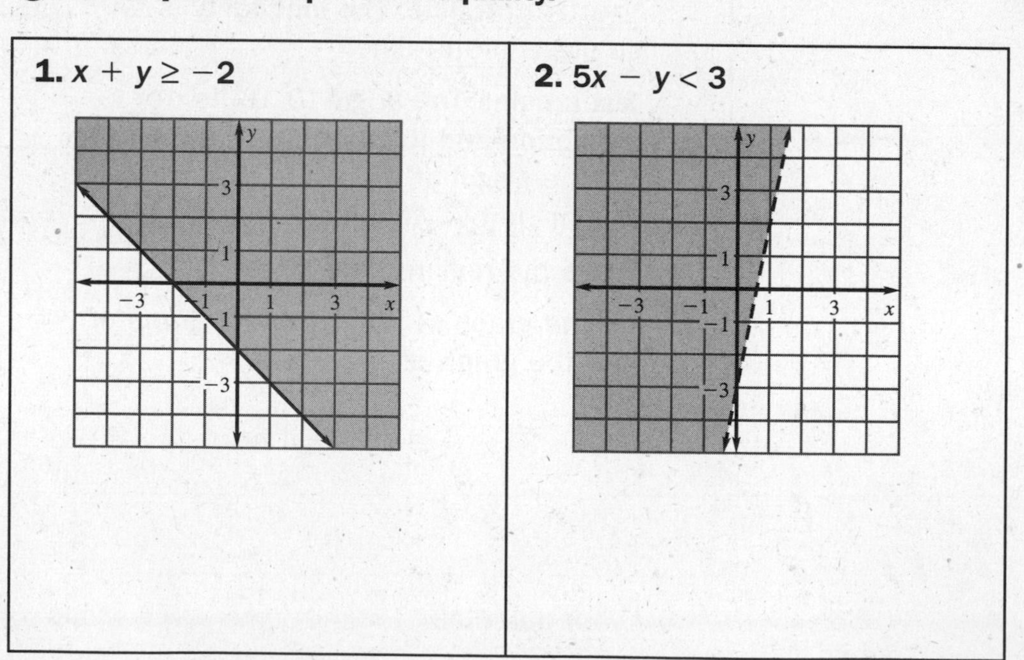

Words to Review

Give an example of the vocabulary word.

Graph of an inequality	**Equivalent inequalities**
$-5\ -4\ -3\ -2\ -1\ 0\ 1\ 2$ $x > -2$	$0 < 2$ and $4 < 6$ are equivalent inequalities.
Compound inequality $-4 \leq x \leq 4$ $x < -4$ or $x > 4$	**Absolute-value equation** $\lvert 4x + 9 \rvert = 17$
Absolute-value inequality $\lvert x - 6 \rvert < 10$	**Linear inequality in two variables** $2x + 3y > 7$

Review your notes and Chapter 6 by using the Chapter Review on pages 375–378 of your textbook.

7.1 Graphing Linear Systems

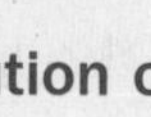

Goal Estimate the solution of a system of linear equations by graphing.

VOCABULARY

System of linear equations A system of linear equations is two or more linear equations in the same variable form. This is also called a linear system.

Solution of a linear system A solution of a linear system is an ordered pair (x, y) that makes each equation in the system a true statement.

Point of intersection A point (a, b) that lies on the graphs of two or more equations is a point of intersection for the graphs.

Example 1 *Find the Point of Intersection*

Use the graph at the right to estimate the solution of the linear system. Then check your solution algebraically.

$x + 2y = -4$ **Equation 1**

$x - 3y = 1$ **Equation 2**

Solution

The lines appear to intersect once at $(\underline{-2}, \underline{-1})$.

Check Substitute $\underline{-2}$ for x and $\underline{-1}$ for y in each equation.

$x + 2y = -4$ $\qquad$ $x - 3y = 1$

$\underline{-2} + 2(\underline{-1}) \stackrel{?}{=} -4$ $\qquad$ $\underline{-2} - 3(\underline{-1}) \stackrel{?}{=} 1$

$\underline{-4} \underline{=} -4$ $\qquad$ $\underline{1} \underline{=} 1$

Answer Because $(\underline{-2}, \underline{-1})$ is a solution of each equation, $(\underline{-2}, \underline{-1})$ is the solution of the system of linear equations.

SOLVING A LINEAR SYSTEM USING GRAPH-AND-CHECK

Step 1 Write each equation in a form that is easy to graph.

Step 2 Graph both equations in the same coordinate plane.

Step 3 Estimate the coordinates of the point of intersection.

Step 4 Check whether the coordinates give a solution by substituting them into each equation of the original linear system.

A line in slope-intercept form, $y = mx + b$, has a slope of m and a y-intercept of b.

Example 2 *Graph and Check a Linear System*

Use the graph-and-check method to solve the linear system.

$5x + 4y = -12$ **Equation 1**

$3x - 4y = -20$ **Equation 2**

1. Write each equation in slope-intercept form.

Equation 1

$5x + 4y = -12$

$4y = -5x - 12$

$y = -\frac{5}{4}x - 3$

Equation 2

$3x - 4y = -20$

$-4y = -3x - 20$

$y = \frac{3}{4}x + 5$

2. Graph both equations.

3. Estimate from the graph that the point of intersection is (−4 , 2).

4. Check whether (−4 , 2) is a solution by substituting −4 for x and 2 for y in each of the original equations.

Equation 1

$5x + 4y = -12$

$5(-4) + 4(2) \stackrel{?}{=} -12$

$-12 = -12$

Equation 2

$3x - 4y = -20$

$3(-4) - 4(2) \stackrel{?}{=} -20$

$-20 = -20$

Answer Because (−4 , 2) is a solution of each equation in the linear system, (−4 , 2) is a solution of the linear system.

✔ ***Checkpoint*** **Use the graph-and-check method to solve the linear system.**

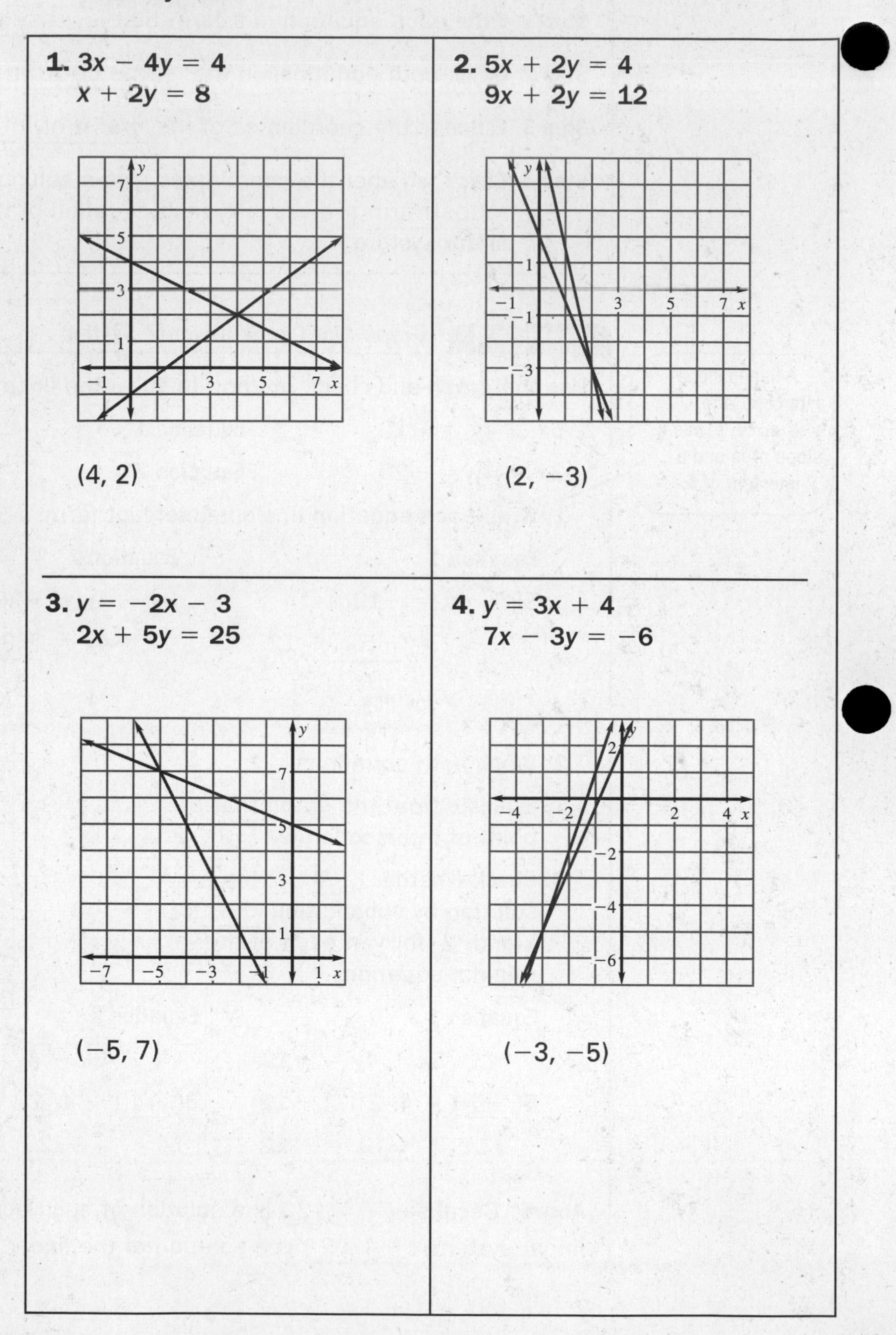

1. $3x - 4y = 4$
$x + 2y = 8$

(4, 2)

2. $5x + 2y = 4$
$9x + 2y = 12$

(2, −3)

3. $y = -2x - 3$
$2x + 5y = 25$

(−5, 7)

4. $y = 3x + 4$
$7x - 3y = -6$

(−3, −5)

7.2 Solving Linear Systems by Substitution

Goal Solve a linear system by substitution.

SOLVING A LINEAR SYSTEM BY SUBSTITUTION

Step 1 **Solve** one of the equations for one of its variables.

Step 2 **Substitute** the expression from Step 1 into the other equation and solve for the other variable.

Step 3 **Substitute** the value from Step 2 into the revised equation from Step 1 and solve.

Step 4 **Check** the solution in each of the original equations.

Example 1 ***Substitution Method: Solve for y First***

Solve the linear system. $4x + y = -5$ **Equation 1**

$3x - y = 5$ **Equation 2**

1. **Solve** for y in Equation 1.

$4x + y = -5$ **Original Equation 1**

$y = -4x - 5$ **Revised Equation 1**

2. **Substitute** $-4x - 5$ for y in Equation 2 and find the value of x.

$3x - y = 5$ **Write Equation 2.**

$3x - (-4x - 5) = 5$ **Substitute $-4x - 5$ for y.**

$7x + 5 = 5$ **Simplify.**

$7x = 0$ **Subtract 5 from each side.**

$x = 0$ **Divide each side by 7.**

3. **Substitute** 0 for x in the revised Equation 1 and find the value of y.

$y = -4x - 5 = -4(0) - 5 = -5$

4. ***Check*** that $(0, -5)$ is a solution by substituting 0 for x and -5 for y in each of the original equations.

When you use the substitution method, you can check the solution by substituting it for x and for y in each of the original equations. You can also use a graph to check your solution.

When using substitution, you will get the same solution whether you solve for y first or x first. You should begin by solving for the variable that is easier to isolate.

Example 2 *Substitution Method: Solve for x First*

Solve the linear system.

$2x - 5y = -13$ **Equation 1**

$x + 3y = -1$ **Equation 2**

Solution

1. Solve for x in Equation 2.

$x + 3y = -1$ **Original Equation 2**

$x = -3y - 1$ **Revised Equation 2**

2. Substitute $-3y - 1$ for x in Equation 1 and find the value of y.

$2x - 5y = -13$ **Write Equation 1.**

$2(-3y - 1) - 5y = -13$ **Substitute $-3y - 1$ for x.**

$-6y - 2 - 5y = -13$ **Use the distributive property.**

$-11y - 2 = -13$ **Combine like terms.**

$-11y = -11$ **Add 2 to each side.**

$y = 1$ **Divide each side by -11.**

3. Substitute 1 for y in the revised Equation 2 and find the value of x.

$x = -3y - 1$ **Write revised Equation 2.**

$x = -3(1) - 1$ **Substitute 1 for y.**

$x = -4$ **Simplify.**

4. *Check* that $(-4, 1)$ is a solution by substituting -4 for x and 1 for y in each of the original equations.

Answer The solution is $(-4, 1)$.

✔ ***Checkpoint*** **Name the variable that you would solve for first. Explain.**

1. $x - 2y = 0$ $x - 8y = -5$ I would isolate x in the first equation because it is easy to do so and the value is easy to substitute in the other equation.	**2.** $4x + 2y = 10$ $7x - y = 12$ I would isolate y in the second equation because it is easier to do so than to isolate y in the first equation or x in either equation.

Use substitution to solve the linear system.

3. $y = x - 1$ $x - 5y = -15$ (5, 4)	**4.** $y = -5x + 3$ $3x + 2y = -8$ (2, −7)

7.3 Solving Linear Systems by Linear Combinations

Goal Solve a system of linear equations by linear combinations.

VOCABULARY

Linear combinations A linear combination of two equations is an equation obtained by (1) multiplying one or both equations by a constant and (2) adding the resulting equations.

SOLVING A LINEAR SYSTEM BY LINEAR COMBINATIONS

Step 1 **Arrange** the equations with like terms in columns.

Step 2 **Multiply**, if necessary, the equations by numbers to obtain coefficients that are opposites for one of the variables.

Step 3 Add the equations from Step 2. Combining like terms with opposite coefficents will eliminate one variable. Solve for the remaining variable .

Step 4 **Substitute** the value obtained from Step 3 into either of the original equations and solve for the other variable .

Step 5 **Check** the solution in each of the original equations.

Example 1 *Add the Equations*

Solve the linear system.

$7x + 2y = -6$ **Equation 1**

$5x - 2y = 6$ **Equation 2**

Solution

Add the equations to get an equation in one variable.

$7x + 2y = -6$	**Write Equation 1.**
$5x - 2y = 6$	**Write Equation 2.**
$\underline{12x} \quad = \underline{0}$	**Add equations.**
$\underline{x} = \underline{0}$	**Solve for $\underline{x}$.**

Substitute $\underline{0}$ for $\underline{x}$ in the first equation and solve for $\underline{y}$.

$7(\underline{0}) + 2y = -6$	**Substitute $\underline{0}$ for $\underline{x}$.**
$\underline{y} = \underline{-3}$	**Solve for $\underline{y}$.**

Check that ($\underline{0}$, $\underline{-3}$) is a solution by substituting $\underline{0}$ for x and $\underline{-3}$ for y in each of the original equations.

Answer The solution is ($\underline{0}$, $\underline{-3}$).

✔ ***Checkpoint*** **Use linear combinations to solve the system of linear equations. Then check your solution.**

1. $4x + y = -4$ $-4x + 2y = 16$	**2.** $4x + 3y = 10$ $12x - 3y = 6$
$(-2, 4)$	$(1, 2)$

Example 2 *Multiply then Add*

Solve the linear system.

$3x - 5y = 15$ **Equation 1**

$2x + 4y = -1$ **Equation 2**

Solution

You can get the coefficients of x to be opposites by multiplying the first equation by 2 and the second equation by -3.

$3x - 5y = 15$ **Multiply by 2.** $\rightarrow$ $6x - 10y = 30$

$2x + 4y = -1$ **Multiply by (-3).** $\rightarrow$ $-6x - 12y = 3$

Add the equations and solve for y.

$$-22y = 33$$
$$y = -1.5$$

Substitute -1.5 for y in the second equation and solve for x.

$2x + 4y = -1$	**Write Equation 2.**
$2x + 4(-1.5) = -1$	**Substitute -1.5 for y.**
$2x - 6 = -1$	**Simplify.**
$x = 2.5$	**Solve for x.**

Answer The solution is $(2.5, -1.5)$.

Checkpoint **Use linear combinations to solve the system of linear equations. Then check your solution.**

3. $x - 3y = 8$ $3x + 4y = 11$ $(5, -1)$	**4.** $6x + 5y = 23$ $9x - 2y = -32$ $(-2, 7)$

7.4 Linear Systems and Problem Solving

Goal Use linear systems to solve real-life problems.

Example 1 *Choosing a Solution Method*

Health Food A health food store mixes granola and raisins to make 20 pounds of raisin granola. Granola costs \$4 per pound and raisins cost \$5 per pound. How many pounds of each should be included for the mixture to cost a total of \$85?

Solution

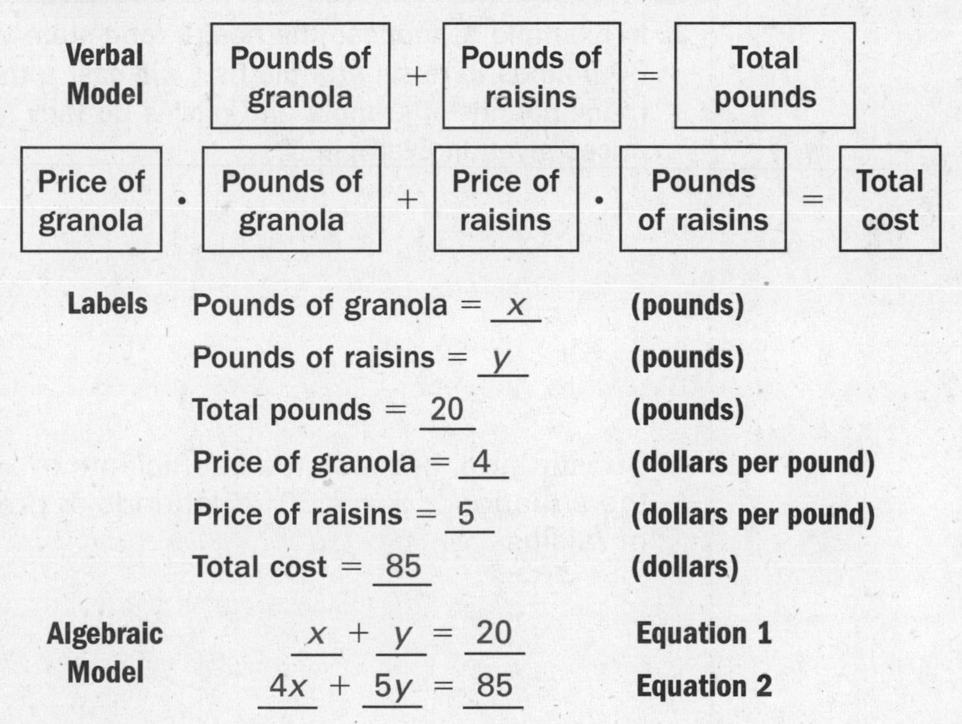

Verbal Model

[Pounds of granola] + [Pounds of raisins] = [Total pounds]

[Price of granola] • [Pounds of granola] + [Price of raisins] • [Pounds of raisins] = [Total cost]

Labels

Pounds of granola =	x	(pounds)
Pounds of raisins =	y	(pounds)
Total pounds =	20	(pounds)
Price of granola =	4	(dollars per pound)
Price of raisins =	5	(dollars per pound)
Total cost =	85	(dollars)

Algebraic Model

$x + y = 20$ **Equation 1**

$4x + 5y = 85$ **Equation 2**

Because the coefficients of x and y are 1 in Equation 1, substitution is most convenient. Solve Equation 1 for x and substitute the result in Equation 2. Simplify to obtain $y = 5$. Substitute 5 for y in Equation 1 and solve for x.

Answer The solution is 5 pounds of raisins and 15 pounds of granola.

WAYS TO SOLVE A SYSTEM OF LINEAR EQUATIONS

Substitution requires that one of the variables be isolated on one side of the equation. It is especially convenient when one of the variables has a coefficient of 1 or -1.

Linear Combinations can be applied to any system, but it is especially convenient when a variable appears in different equations with coefficients that are opposites.

Graphing can provide a useful method for estimating a solution.

Checkpoint **Choose a method to solve the linear system. Explain your choice, and then solve the system.**

1. In Example 1, suppose the health food store wants to make 30 pounds of raisin granola that will cost a total of $125. How many pounds of granola and raisins do they need? Use the prices given in Example 1.

Substitution, because the coefficients of x and y are 1 in the equation $x + y = 30$; 25 pounds of granola, 5 pounds of raisins

7.5 Special Types of Linear Systems

Goal Identify how many solutions a linear system has.

NUMBER OF SOLUTIONS OF A LINEAR SYSTEM

If the two solutions have different **slopes, then the system has one solution.**

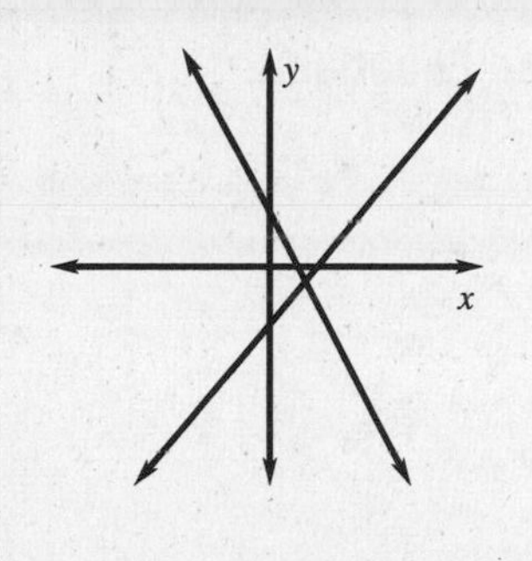

Lines intersect:

Exactly one **solution.**

If the two solutions have the same **slope but** different ***y*-intercepts, then the system has no solution.**

Lines are parallel:

No **solution.**

If the two equations have the same **slope and the** same ***y*-intercepts, then the system has infinitely many solutions.**

Lines coincide:

Infinitely many **solutions.**

Example 1 *A Linear System with No Solution*

Show that the linear system has no solution.

$-x + y = -3$ **Equation 1**

$x + y = 2$ **Equation 2**

Solution

Method 1: Graphing Rewrite each equation in slope-intercept form. Then graph the linear system.

$y = \underline{x - 3}$ **Revised Equation 1**

$y = \underline{x + 2}$ **Revised Equation 2**

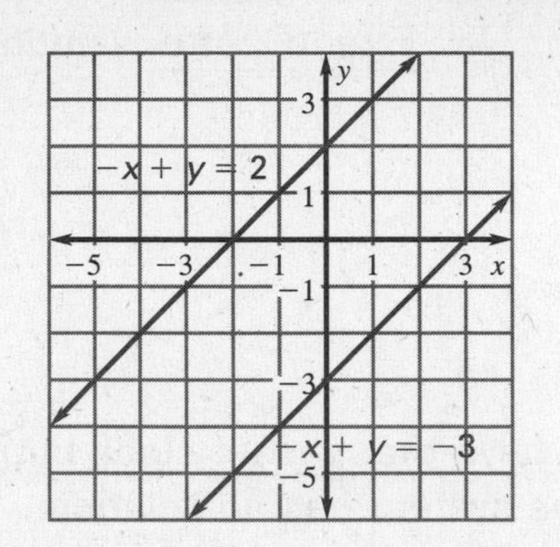

Answer Because the lines have the same slope but different *y*-intercepts, they are parallel. Parallel lines do not intersect, so the system has no solution.

Method 2: Substitution Because Equation 2 can be rewritten as $y = \underline{x + 2}$, you can substitute $\underline{x + 2}$ for *y* in Equation 1.

$-x + y = -3$ **Write Equation 1.**

$-x + \underline{x + 2} = -3$ **Substitute $\underline{x + 2}$ for *y*.**

$2 \neq -3$ **Combine like terms.**

Answer The variables are eliminated and you are left with a statement that is false regardless of the values of *x* and *y*. This tells you that the system has no solution.

Example 2 *A Linear System with Infinitely Many Solutions*

Show that the linear system has many solutions.

$3x + y = -1$ **Equation 1**

$-6x - 2y = 2$ **Equation 2**

Solution

Method 1: Graphing Rewrite each equation in slope-intercept form. Then graph the linear system.

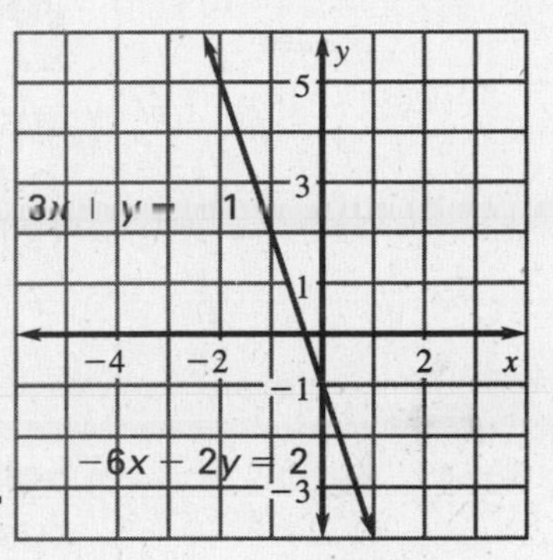

$y = \underline{-3x - 1}$ **Revised Equation 1**

$y = \underline{-3x - 1}$ **Revised Equation 2**

Answer From these equations you can see that the equations represent the same line. ___Every___ point on the line is a solution.

Method 2: Linear Combinations You can multiply Equation 1 by ___2___.

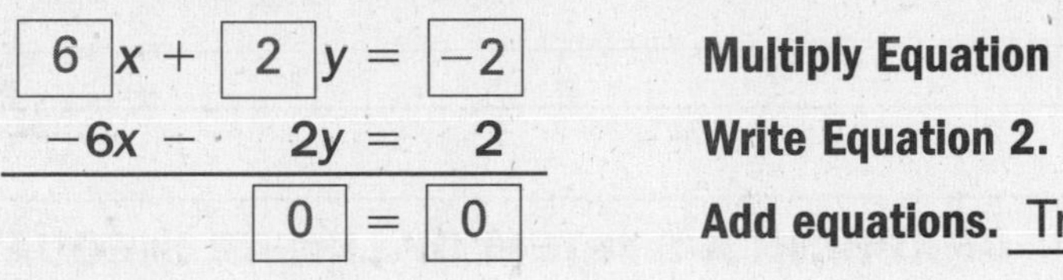

$\boxed{6}x + \boxed{2}y = \boxed{-2}$ **Multiply Equation 1 by ___2___.**

$-6x - 2y = 2$ **Write Equation 2.**

$\boxed{0} = \boxed{0}$ **Add equations. ___True___ statement**

Answer The variables are ___eliminated___ and you are left with a statement that is ___true___ regardless of the values of x and y. This tells you that the system has ___infinitely many solutions___.

✔ ***Checkpoint*** **Solve the linear system and tell how many solutions the system has.**

1. $x - 2y = 3$ $-5x + 10y = -15$ infinitely many solutions	**2.** $-2x + 3y = 4$ $-4x + 6y = 10$ no solution

7.6 Systems of Linear Inequalities

Goal Graph a system of linear inequalities.

VOCABULARY

System of linear inequalities Two or more linear inequalities in the same variables form a system of linear inequalities. This is also called a system of inequalities.

Solution of a system of linear inequalities A solution of a system of linear inequalities is an ordered pair that is a solution of each inequality in the system.

GRAPHING A SYSTEM OF LINEAR INEQUALITIES

Step 1 Graph **the boundary lines of each inequality. Use a** dashed **line if the inequality is $<$ or $>$ and a** solid **line if the inequality is $\leq$ or $\geq$.**

Step 2 Shade **the appropriate half-plane for each inequality.**

Step 3 Identify **the solution of the system of inequalities as the intersection of the half-planes from Step 2.**

Example 1 *Graph a System of Two Linear Inequalities*

Graph the system of linear inequalities.

$y - x \geq -1$ **Inequality 1**

$x + 2y < 1$ **Inequality 2**

To check your graph, choose a point in the overlap of the half-planes. Then substitute the coordinates into each inequality. If each inequality is true, then the point is a solution.

Solution

Graph both inequalities in the same coordinate plane. The graph of the system is the overlap, or intersection , of the two half-planes.

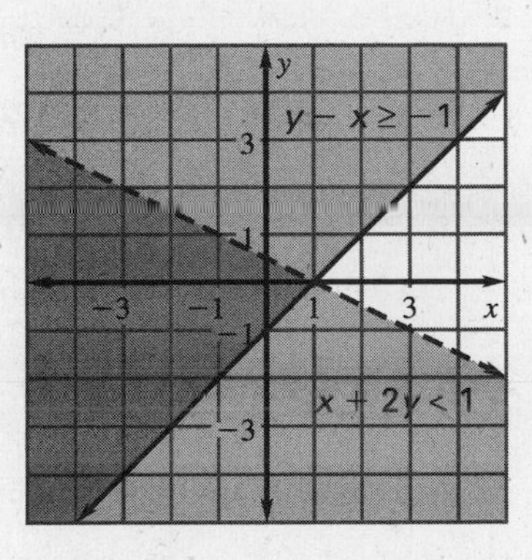

Example 2 *Graph a System of Three Linear Inequalities*

Graph the system of linear inequalities.

$y \geq -3$ **Inequality 1**

$x < 2$ **Inequality 2**

$y < x + 1$ **Inequality 3**

Solution

The graph of $y \geq -3$ is the half-plane on and above the solid line $y = -3$.

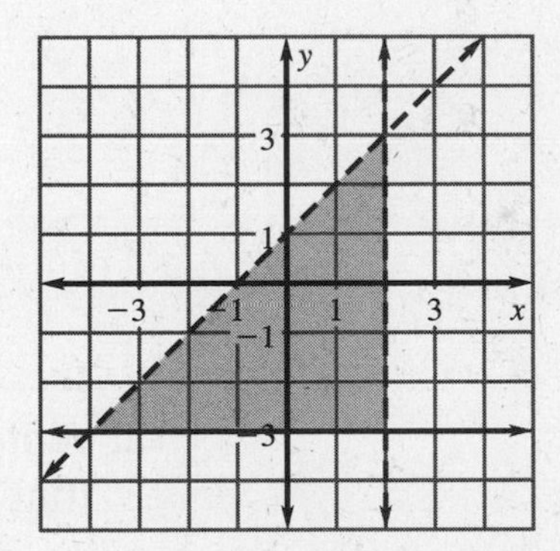

The graph of $x < 2$ is the half-plane to the left of the dashed line $x = 2$.

The graph of $y < x + 1$ is the half-plane below the dashed line $y = x + 1$.

Finally, the graph of the system is the overlap, or intersection, of the three half-planes.

Example 3 *Write a System of Linear Inequalities*

Write a system of inequalities that defines the shaded region at the right.

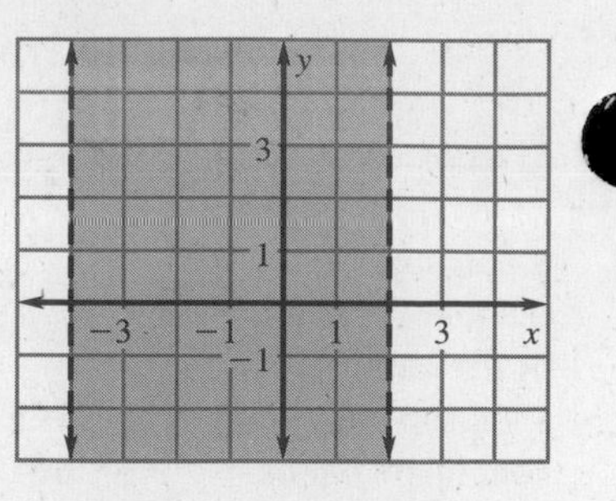

Solution

The graph of one inequality is the half-plane to the *left* of $x = 2$.

The graph of the other inequality is the half-plane to the *right* of $x = -4$.

The shaded region of the graph is the vertical band that lies between the two vertical lines, $x = 2$ and $x = -4$, but not on the lines.

Answer The system of linear inequalities below defines the shaded region.

$x < 2$ **Inequality 1**

$x > -4$ **Inequality 2**

Checkpoint Complete the following exercises.

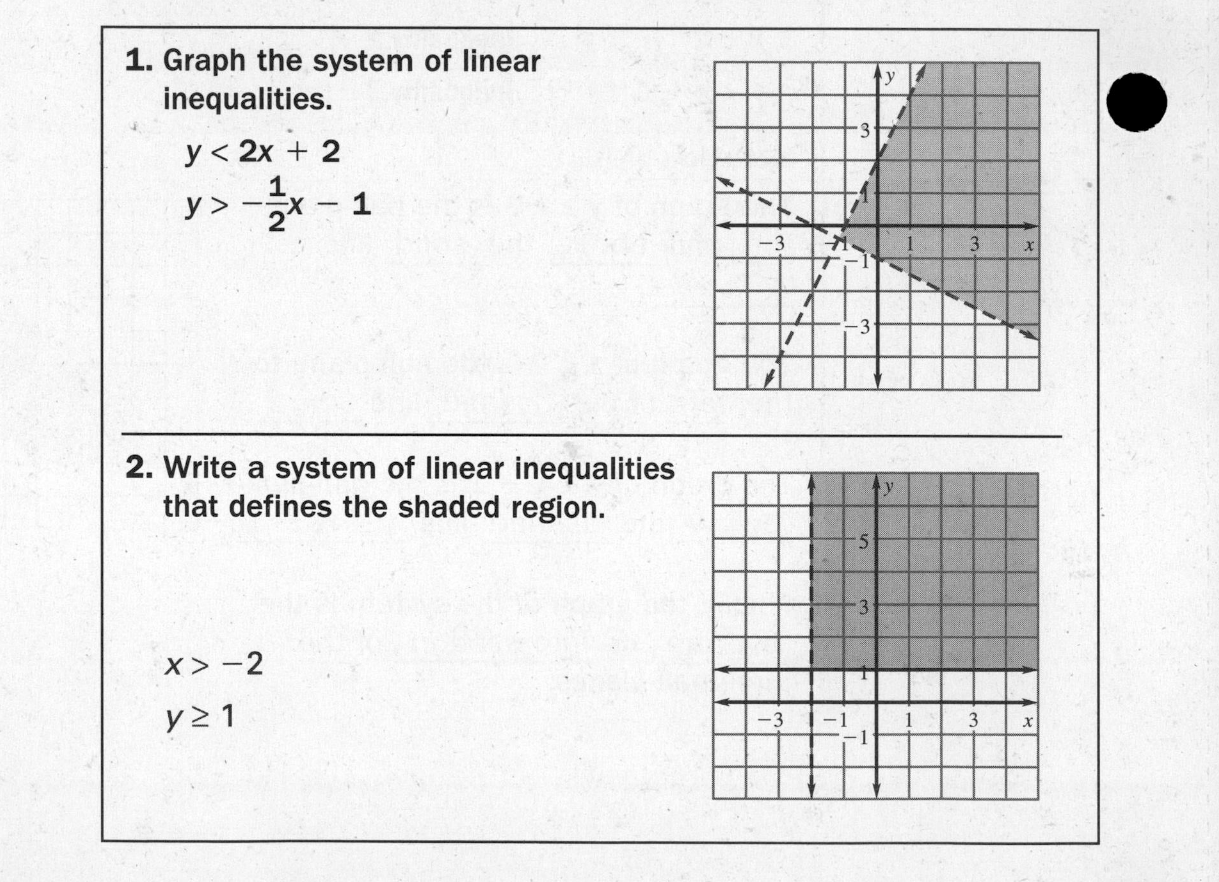

1. Graph the system of linear inequalities.

$y < 2x + 2$

$y > -\frac{1}{2}x - 1$

2. Write a system of linear inequalities that defines the shaded region.

$x > -2$

$y \geq 1$

Words to Review

Give an example of the vocabulary word.

System of linear equations	**Solution of a linear system**
$y = -x + 4$ $y = x - 4$	The solution of the system $y = x - 4$ $y = -x + 4$ is (4, 0).
Point of intersection	**Linear combination**
The point of intersection of the system $y = x - 4$ $y = -x + 4$ is (4, 0).	$12x = 0$ is a linear combination of the two equations below. $7x + 2y = -6$ $5x - 2y = 6$ $12x = 0$
System of linear inequalities	**Solution of a system of linear inequalities**
$x > 2$ $y \leq -x + 4$ $y > x - 4$	A solution of the system: $y \geq -3$ $x < 2$ $y < x + 1$ is (0, 0).

Review your notes and Chapter 7 by using the Chapter Review on pages 431–434 of your textbook.

8.1 Multiplication Properties of Exponents

Goal Use multiplication properties of exponents.

MULTIPLICATION PROPERTIES OF EXPONENTS

Let a and b be numbers and let m and n be positive integers.

Product of Powers Property

To multiply powers having the same base, add the exponents.

$a^m \cdot a^n = a^{m+n}$

Power of a Power Property

To find a power of a power, multiply the exponents.

$(a^m)^n = a^{m \cdot n}$

Power of a Product Property

To find a power of a product, find the power of each factor and multiply.

$(a \cdot b)^m = a^m \cdot b^m$

Example 1 ***Use the Product of Powers Property***

Write the expression as a single power of the base.

a. $4^3 \cdot 4^5 = 4^{3+5}$ — **Use product of powers property.**

$= 4^8$ — Add **the exponents.**

b. $(-3)(-3)^7 = (-3)^1 \cdot (-3)^7$ — **Rewrite (-3) as $(-3)^1$.**

$= (-3)^{1+7}$ — **Use** product of powers **property.**

$= (-3)^8$ — Add the exponents.

c. $y^4 \cdot y^5 \cdot y^6 = y^{4+5+6}$ — Use product of powers property.

$= y^{15}$ — Add the exponents.

Example 2 ***Use the Power of a Power Property***

Write the expression as a single power of the base.

a. $(7^3)^5 = 7^{\underline{3 \cdot 5}}$ **Use** power of a power **property.**

$= 7^{\underline{15}}$ Multiply **exponents.**

b. $(x^2)^5 = x^{\underline{2 \cdot 5}}$ Use power of a power property.

$= \underline{x^{10}}$ Multiply exponents.

Checkpoint **Simplify the expression.**

1. $(-4)^3 \cdot (-4)^6$	**2.** $w^{10} \cdot w$	**3.** $m^4 \cdot m^2 \cdot m^5$
$(-4)^9$	w^{11}	m^{11}

Write the expression as a single power of the base.

4. $(6^4)^5$	**5.** $[(-5)^3]^6$	**6.** $(n^9)^3$
6^{20}	$(-5)^{18}$	n^{27}

Example 3 ***Use the Power of a Product Property***

Simplify the expression.

a. $(2 \cdot 3)^3 = 2^{\underline{3}} \cdot 3^{\underline{3}}$ **Use** power of a product **property.**

$= \underline{8} \cdot \underline{27}$ **Evaluate each** power**.**

$= \underline{216}$ **Multiply.**

b. $(9xy)^2 = 9^{\underline{2}} \cdot x^{\underline{2}} \cdot y^{\underline{2}}$ Use power of a product property.

$= \underline{81x^2y^2}$ **Evaluate power.**

Example 4 ***Use All Three Properties***

Simplify the expression $(7z^3)^2 \cdot z^4$.

Solution

$(7z^3)^2 \cdot z^4 = 7^{\underline{2}} \cdot (z^3)^{\underline{2}} \cdot z^4$ — **Use** power of a product **property.**

$= \underline{49} \cdot z^{\underline{6}} \cdot z^4$ — **Use** power of a power **property.**

$= \underline{49z^{10}}$ — **Use** product of powers **property.**

Checkpoint **Simplify the expression.**

7. $(-4x^2y^2)^2$	**8.** $(6p)^2$	**9.** $(-2xy)^4$
$16x^4y^4$	$36p^2$	$16x^4y^4$
10. $(3a^2)^3$	**11.** $(x^5)^3 \cdot x^7$	**12.** $(2wz^2)^5(wz)^2$
$27a^6$	x^{22}	$32w^7z^{12}$

8.2 Zero and Negative Exponents

Goal Evaluate powers that have zero or negative exponents.

ZERO AND NEGATIVE EXPONENTS

Let a be a nonzero number and let n be an integer.

- A nonzero number to the zero power is 1: $a^0 = 1, a \neq 0$
- a^{-n} is the reciprocal of a^n: $a^{-n} = \frac{1}{a^n}, a \neq 0$

Example 1 ***Powers with Zero and Negative Exponents***

Evaluate the expression.

a. $(21)^0 = \underline{1}$

b. $\left(\frac{2}{5}\right)^0 = \underline{1}$

c. $4^{-2} = \underline{\frac{1}{4^2}} = \underline{\frac{1}{16}}$

d. $\frac{1}{3^{-3}} = \underline{3^3} = \underline{27}$

Checkpoint Evaluate the expression.

1. 2^0	**2.** $(-1)^0$	**3.** 2^{-3}	**4.** $\frac{1}{(-8)^{-2}}$
1	1	$\frac{1}{8}$	64

Example 2 *Evaluate Exponential Expressions*

Evaluate the expression.

a. $2^4 \cdot 2^{-4}$ **b.** $(3^{-2})^{-2}$ **c.** $(-2 \cdot 3)^{-2}$

Solution

a. $2^4 \cdot 2^{-4} = 2^{\underline{4 + (-4)}}$ Use product of powers property.

$= 2^{\underline{0}}$ Add exponents.

$= \underline{1}$ $a^{\underline{0}}$ is equal to $\underline{1}$.

b. $(3^{-2})^{-2} = 3^{\underline{-2 \cdot (-2)}}$ Use power of a power property.

$= 3^{\underline{4}}$ Multiply exponents.

$= \underline{81}$ Evaluate power.

c. $(-2 \cdot 3)^{-2} = \dfrac{\boxed{1}}{\boxed{(-2 \cdot 3)^2}}$ Definition of negative exponent

$= \dfrac{\boxed{1}}{\boxed{(-2)^2 \cdot 3^2}}$ Use power of a product property.

$= \dfrac{\boxed{1}}{\boxed{4 \cdot 9}}$ Evaluate powers.

$= \dfrac{1}{36}$ Simplify.

✔ ***Checkpoint*** **Evaluate the expression without using a calculator.**

5. $7^4 \cdot 7^{-4}$	**6.** $(2^{-2})^{-2}$	**7.** $(-3 \cdot 3)^{-2}$
1	16	$\frac{1}{81}$

Example 3 *Simplify Exponential Expressions*

Rewrite the expression with positive exponents.

a. $3x^{-4}y^{-5}$ **b.** $\frac{a^{-1}}{b^{-2}}$ **c.** $(7c)^{-2}$

Solution

a. $3x^{-4}y^{-5} = 3 \cdot \frac{1}{x^4} \cdot \frac{1}{y^5}$ **Definition of negative exponents**

$= \frac{3}{x^4y^5}$ **Multiply.**

b. $\frac{a^{-1}}{b^{-2}} = a^{-1} \cdot \frac{1}{b^{-2}}$ **Multiply by reciprocal.**

$= \frac{1}{a} \cdot b^2$ **Definition of negative exponents**

$= \frac{b^2}{a}$ **Multiply.**

c. $(7c)^{-2} = \frac{1}{(7c)^2}$ **Definition of negative exponents**

$= \frac{1}{7^2 \cdot c^2}$ **Use** power of a product **property.**

$= \frac{1}{49c^2}$ **Evaluate power.**

✓ ***Checkpoint*** **Rewrite the expression with positive exponents.**

8. $3c^7d^{-7}$	**9.** $\frac{u^{-5}}{v^{-3}}$	**10.** $(3t)^{-3}$
$\frac{3c^7}{d^7}$	$\frac{v^3}{u^5}$	$\frac{1}{27t^3}$

8.3 Graphs of Exponential Functions

Goal Graph an exponential function.

VOCABULARY

Exponential function An exponential function is a function of the form $y = ab^x$, where $b > 0$, and $b \neq 1$.

Example 1 ***Graph an Exponential Function when b > 1***

Graph the function $y = 3^x$.

Solution

Make a table of values that includes both positive and negative *x*-values.

x	−2	−1	0	1	2
y	$3^{-2} = \frac{1}{9}$	$3^{-1} = \frac{1}{3}$	$3^0 = 1$	$3^1 = 3$	$3^2 = 9$

Write the coordinates of the five points given by the table:

$\left(-2, \frac{1}{9}\right), \left(-1, \frac{1}{3}\right), (0, 1), (1, 3), (2, 9)$

Draw a coordinate plane and plot the five points listed above. Then draw a smooth curve through the points.

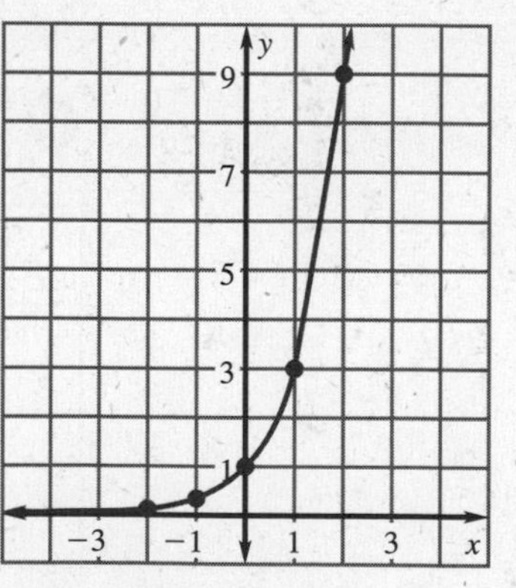

Notice that the graph has a *y*-intercept of 1, and that it gets closer to the negative side of the *x*-axis as the *x*-values decrease.

Example 2 *Graph an Exponential Function when $0 < b < 1$*

Graph the function $y = 2\left(\frac{1}{2}\right)^x$.

Solution Make a table of values.

Write the coordinates of the five points given by the table:

> Be sure to follow the order of operations when evaluating the function.

x	-2	-1	0	1	2
y	$2\left(\frac{1}{2}\right)^{-2} = \underline{8}$	$2\left(\frac{1}{2}\right)^{-1} = \underline{4}$	$2\left(\frac{1}{2}\right)^{0} = \underline{2}$	$2\left(\frac{1}{2}\right)^{1} = \underline{1}$	$2\left(\frac{1}{2}\right)^{2} = \underline{\frac{1}{2}}$

Draw a coordinate plane and plot the five points given by the table. Then draw a smooth curve through the points.

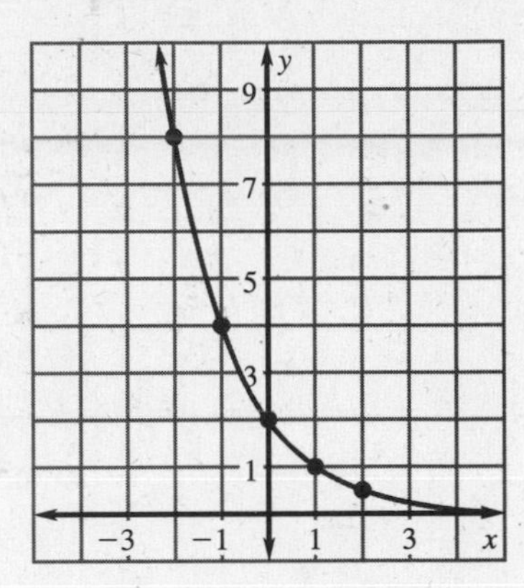

Notice that the graph has a y-intercept of <u>2</u>, and that it gets closer to the <u>positive</u> side of the x-axis as the x-values <u>increase</u>.

Example 3 *Find Domain and Range*

Find the domain and range of the function.

a. $y = 3^x$

b. $y = 2\left(\frac{1}{2}\right)^x$

Solution

a. You can see from the graph of the function in Example 1 that $y = 3^x$ is defined for <u>all x-values</u>, but only has y-values greater than <u>0</u>. So the domain of $y = 3^x$ is <u>all real numbers</u> and the range is <u>all positive real numbers</u>.

b. You can see from the graph of the function in Example 2 that $y = 2\left(\frac{1}{2}\right)^x$ is defined for <u>all x-values</u>, but only has y-values greater than <u>0</u>. So the domain of $y = 2\left(\frac{1}{2}\right)^x$ is <u>all real numbers</u> and the range is <u>all positive real numbers</u>.

Checkpoint Complete the following exercises.

1. Make a table of values and graph the function $y = \frac{1}{2}(2)^x$. Use the x-values of -1, 0, 1, 2, 3, and 4.

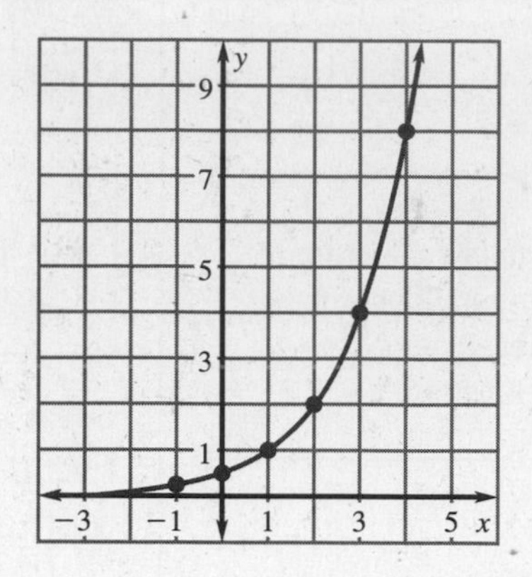

x	-1	0	1	2	3	4
y	$\frac{1}{4}$	$\frac{1}{2}$	1	2	4	8

2. Describe the domain and the range of $y = \frac{1}{2}(2)^x$.

Domain: all real numbers; Range: all positive real numbers

8.4 Division Properties of Exponents

Goal Use division properties of exponents.

DIVISION PROPERTIES OF EXPONENTS

Let a and b be real numbers and let m and n be integers.

Quotient of Powers Property

To divide powers that have the same base, subtract the exponents.

$$\frac{a^m}{a^n} = a^{m-n}, a \neq 0$$

Power of a Quotient Property

To find a power of a quotient, find the power of the numerator and the power of the denominator and divide.

$$\left(\frac{a}{b}\right)^m = \frac{a^m}{b^m}, b \neq 0$$

Example 1 ***Use the Quotient of Powers Property***

Simplify the quotient.

a. $\frac{8^4}{8^3} = 8^{4-3}$ **Use quotient of powers property.**

$= 8^1$ Subtract **exponents.**

$= 8$ **Evaluate power.**

b. $\frac{x^4}{x^7} = x^{4-7}$ **Use quotient of powers property.**

$= x^{-3}$ Subtract **exponents.**

$= \frac{1}{x^3}$ **Use definition of negative exponent.**

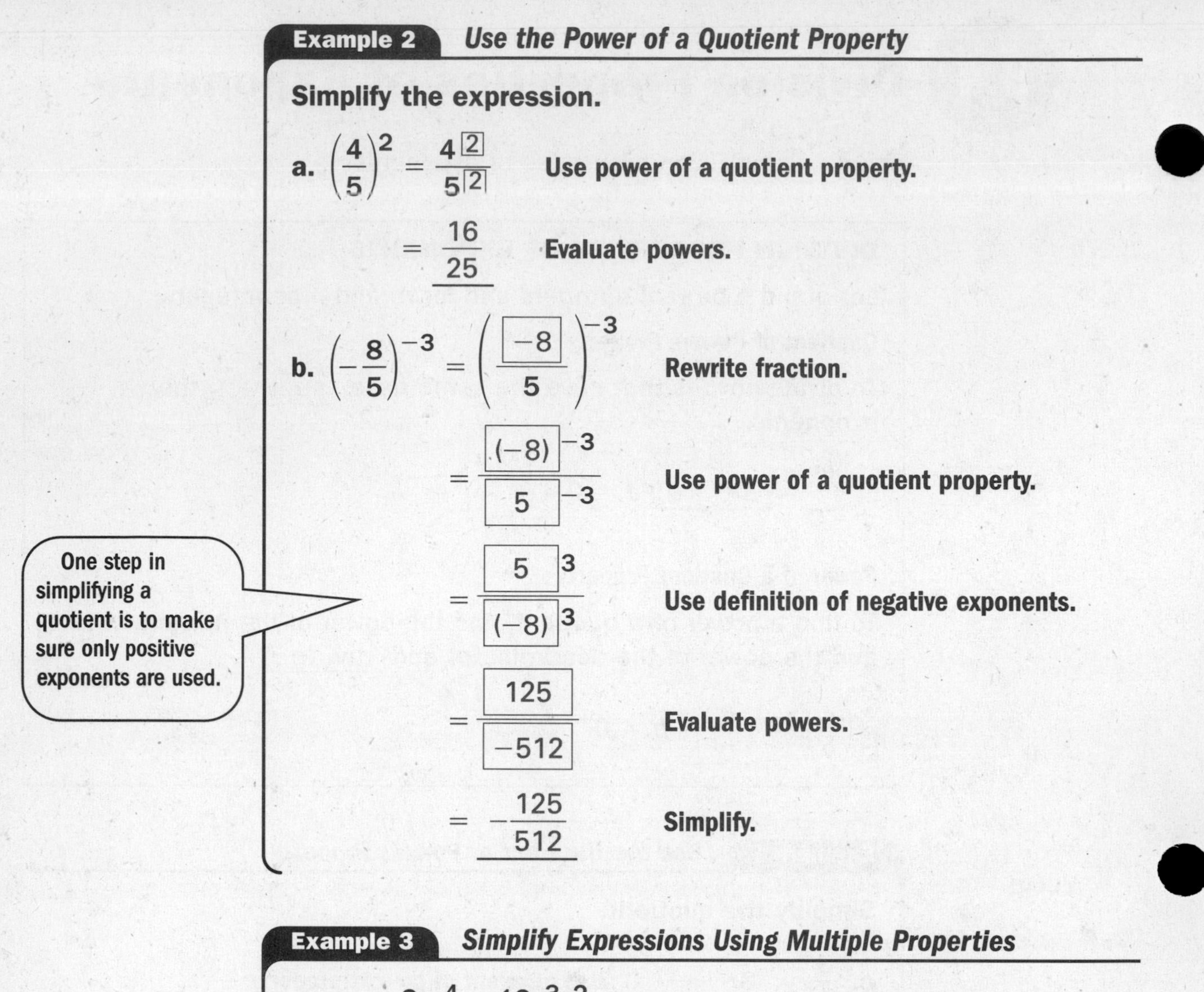

Example 2 *Use the Power of a Quotient Property*

Simplify the expression.

a. $\left(\frac{4}{5}\right)^2 = \frac{4^{\boxed{2}}}{5^{\boxed{2}}}$ **Use power of a quotient property.**

$= \frac{16}{25}$ **Evaluate powers.**

b. $\left(-\frac{8}{5}\right)^{-3} = \left(\frac{\boxed{-8}}{5}\right)^{-3}$ **Rewrite fraction.**

$= \frac{\boxed{(-8)}^{-3}}{\boxed{5}^{-3}}$ **Use power of a quotient property.**

$= \frac{\boxed{5}^{3}}{\boxed{(-8)}^{3}}$ **Use definition of negative exponents.**

$= \frac{\boxed{125}}{\boxed{-512}}$ **Evaluate powers.**

$= -\frac{125}{512}$ **Simplify.**

One step in simplifying a quotient is to make sure only positive exponents are used.

Example 3 *Simplify Expressions Using Multiple Properties*

Simplify $\frac{3xy^4}{4x^3} \cdot \frac{12x^3y^2}{x^2}$. Use only positive exponents.

Solution

$\frac{3xy^4}{4x^3} \cdot \frac{12x^3y^2}{x^2} = \frac{\boxed{36x^4y^6}}{\boxed{4x^5}}$ **Use product of powers property.**

$= 9x^{-1}y^6$ **Use quotient of powers property.**

$= \frac{9y^6}{x}$ **Use definition of negative exponents.**

Example 4 *Simplify Expressions with Negative Exponents*

Simplify the expression $\frac{b}{a^{-3}} \cdot \left(\frac{b^3}{a}\right)^{-2}$. Use positive exponents.

Solution

$$\frac{b}{a^{-3}} \cdot \left(\frac{b^3}{a}\right)^{-2} = \frac{b}{a^{-3}} \cdot \frac{(b^3)^{-2}}{a^{-2}}$$ **Use power of a quotient property.**

$$= b \cdot a^3 \cdot \frac{a^2}{(b^3)^2}$$ **Use definition of negative exponents.**

$$= \frac{ba^5}{b^6}$$ **Use product of powers property.** **Use power of a power property.**

$$= b^{-5}a^5$$ **Use quotient of powers property.**

$$= \frac{a^5}{b^5}$$ **Use definition of negative exponents.**

✔ ***Checkpoint*** **Simplify the expression. Write any fractions in simplest form using only positive exponents.**

1. $\frac{4^6}{4^4}$ 16	2. $\frac{t^7}{t^8}$ $\frac{1}{t}$	3. $\left(\frac{-5}{c}\right)^2$ $\frac{25}{c^2}$
4. $\left(\frac{3}{4}\right)^{-2}$ $\frac{16}{9}$	5. $\frac{5x^2y^3}{6x^4} \cdot \frac{24x^5y^2}{x^6y^3}$ $\frac{20y^2}{x^3}$	6. $\frac{a^3}{b^{-4}} \cdot \left(\frac{a}{b^2}\right)^{-3}$ b^{10}

8.5 Scientific Notation

Goal Read and write numbers in scientific notation.

VOCABULARY

Scientific notation A number is written in scientific notation if it is of the form $c \times 10^n$, where $1 \leq c < 10$ and n is an integer.

Example 1 ***Write Numbers in Decimal Form***

Write the number in decimal form.

a. $3.69 \times 10^1 =$ 36.9 — **Move decimal point** 1 **place to the** right**.**

b. $7.3 \times 10^4 =$ 73,000 — **Move decimal point** 4 **places to the** right**.**

c. $9 \times 10^{-2} =$ 0.09 — **Move decimal point** 2 **places to the** left**.**

d. $4.5 \times 10^{-3} =$ 0.0045 — **Move decimal point** 3 **places to the** left**.**

Example 2 ***Write Numbers in Scientific Notation***

Write the number in scientific notation.

a. $618 = 6.18 \times 10^{2}$ — **Move decimal point** 2 **places to the** left**.**

b. $2.5 = 2.5 \times 10^{0}$ — **Move decimal point** 0 **places.**

c. $0.0098 = 9.8 \times 10^{-3}$ — **Move decimal point** 3 **places to the** right**.**

 ***Checkpoint*** **Write the number in decimal form.**

1. 2.49×10^5	**2.** 5.6×10^{-4}	**3.** 1.0001×10^4
249,000	0.00056	10,001

Write the number in scientific notation.

4. 0.006	**5.** 82,000,000	**6.** 0.00037
6×10^{-3}	8.2×10^7	3.7×10^{-4}

Example 3 ***Operations with Scientific Notation***

Perform the indicated operation. Write the result in scientific notation.

a. $\frac{5.4 \times 10^{-2}}{7.2 \times 10^{-6}} = \frac{5.4}{7.2} \times \frac{10^{-2}}{10^{-6}}$ — **Write as a product.**

$= \underline{0.75} \times 10^{\underline{4}}$ — **Use quotient of powers property.**

$= (\underline{7.5} \times 10^{\underline{-1}}) \times 10^{\underline{4}}$ — **Write in scientific notation.**

$= \underline{7.5} \times 10^{\underline{3}}$ — **Use product of powers property.**

b. $(3.0 \times 10^{-4})^3 = \underline{3^3} \times \underline{(10^{-4})^3}$ — **Use power of a product property.**

$= \underline{27} \times 10^{\underline{-12}}$ — **Use power of a power property.**

$= (\underline{2.7} \times 10^{\underline{1}}) \times 10^{\underline{-12}}$ — **Write in scientific notation.**

$= \underline{2.7} \times 10^{\underline{-11}}$ — **Use product of powers property.**

8.6 Exponential Growth Functions

Goal Write and graph exponential growth functions.

VOCABULARY

Exponential growth A quantity is growing exponentially if it is increasing by the same percent r in each unit of time t. Exponential growth can be modeled by the equation $y - C(1 + r)^t$, where C is the initial quantity and y is the amount after t units of time.

Growth rate In an exponential growth model, the growth rate is the proportion by which the quantity increases each unit of time.

Growth factor The growth factor is the expression $(1 + r)$ in the exponential growth model where r is the growth rate.

Example 1 ***Write an Exponential Growth Model***

Property Value Growth The value of a $100,000 house is expected to increase 1% each year over the next fifteen years. Write a model for the expected value of the house during the fifteen years.

Solution

Let y be the value of the house during the fifteen years and let t be the number of years. The initial value of the house C is \$ 100,000. The growth rate r is 1 % or 0.01.

$y = C(1 + r)^t$ **Write exponential growth model.**

$= 100{,}000(1 + 0.01)^t$ **Substitute 100,000 for C, and 0.01 for r.**

$= 100{,}000(1.01)^t$ **Add.**

Example 2 ***Find the Balance in an Account***

Compound Interest You deposit \$450 in an account that pays 2% interest compounded yearly. What will the account balance be after 10 years?

Solution

The initial amount P is \$ 450 , the growth rate is 2 %, and the time is 10 years.

> The model for compound interest is generally written using A (for the account balance) instead of y, and P (for the principal) instead of C.

$A = P(1 + r)^t$ — **Write yearly compound interest model.**

$= 450(1 + 0.02)^{10}$ — **Substitute** 450 **for** P, 0.02 **for** r, **and** 10 **for** t.

$= 450(1.02)^{10}$ — **Add.**

≈ 549 — **Use a calculator.**

Answer The balance after 10 years will be about \$ 549 .

✔ ***Checkpoint*** **Complete the following exercises.**

1. A company with 60 employees expects a 6% yearly increase in the number of employees. Write an exponential growth model to represent the number of employees E after t years.

$E = 60(1.06)^t$

2. You deposit \$375 in an account that pays 3% interest compounded yearly. What is the account balance after 8 years?

about \$475

Example 3 *Use an Exponential Growth Model*

Population Growth An initial population of 80 rabbits doubles each year for four years. What is the rabbit population after four years?

Solution

You know that the population doubles each year. This tells you the factor by which the population is growing, not the percent change in the population. Therefore, the *growth factor* is 2. The initial population is 80 and the time is 4 years.

$y = C(1 + r)^t$	**Write exponential growth model.**
$= 80(2)^4$	**Substitute 80 for C, 2 for $(1 + r)$, and 4 for t.**
$= 1280$	**Evaluate.**

Answer There will be 1280 rabbits after 4 years.

Example 4 *A Model with a Large Growth Rate*

Graph the exponential growth model in Example 3.

Solution

The growth model from Example 3 is $y = 80(2)^t$.

Make a table of values, plot the points in a coordinate plane, and draw a smooth curve through the points.

A large growth rate corresponds to a rapid increase in *y*-values.

t	0	1	2	3	4
y	80	160	320	640	1280

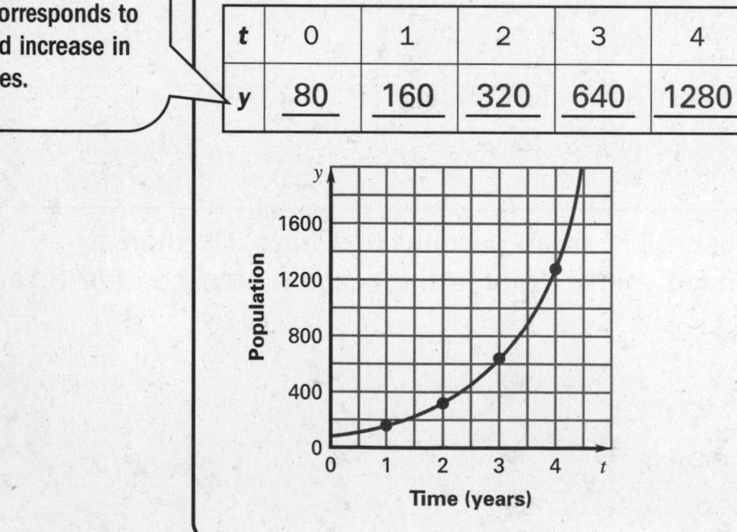

Exponential Decay Functions

Goal Write and graph exponential decay functions.

VOCABULARY

Exponential decay A quantity is decreasing exponentially if it is decreasing by the same percent r in each unit of time t. Exponential decay can be modeled by the equation $y = C(1 - r)^t$, where C is the initial quantity and y is the amount after t units of time.

Decay rate In an exponential decay model, the decay rate is the proportion by which the quantity decreases each unit of time.

Decay factor The decay factor is the expression $(1 - r)$ in the exponential decay model where r is the decay rate.

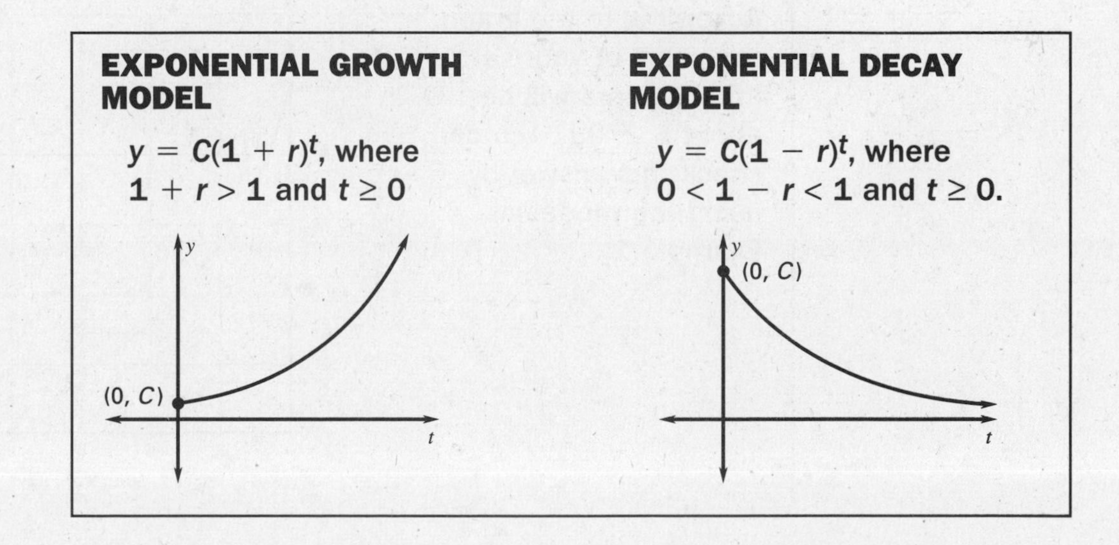

Example 1 *Write an Exponential Decay Model*

Cars A car is purchased for \$20,000. You expect the car to depreciate (lose value) at a rate of 25% per year. Write an exponential decay model to represent this situation. Then find the value of the car after 5 years.

Solution Let y be the value of the car and let t be the number of years of ownership. The initial value of the car C is \$ 20,000 . The decay rate r is 25 % or 0.25 .

$y = C(1 - r)^t$ **Write exponential decay model.**

$y = 20{,}000(1 - 0.25)^t$ **Substitute 20,000 for C and 0.25 for r.**

$y = 20{,}000(0.75)^t$ **Subtract.**

To find the value in 5 years, substitute 5 for t.

$y = 20{,}000(0.75)^t = 20{,}000(0.75)^5 \approx 4746$

Answer The car will be worth about \$ 4746 after 5 years.

Example 2 *Graph an Exponential Decay Model*

Graph the exponential decay model in Example 1. Then use the graph to estimate the value of the car after 7 years.

Solution Make a table of values, plot the points in a coordinate plane, and draw a smooth curve through the points.

t	0	2	4	6	8
y	20,000	11,250	6328	3560	2002

According to the graph, the value of your car after 7 years will be about \$ 2700 . You can check this answer by using the model in Example 1.

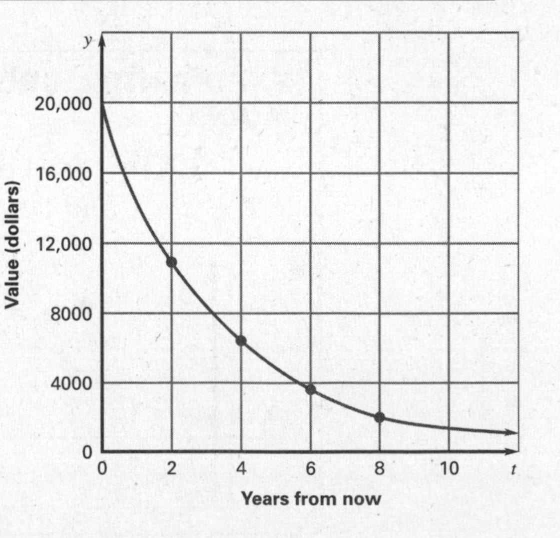

✔ *Checkpoint* **A boat costs \$3200. The value of the boat depreciates at the rate of 13% per year.**

1. Write an exponential decay model to represent this situation.

$y = 3200(0.87)^t$

2. Use the model to find the value of the boat after 3 years.

about \$2107

3. Graph the exponential decay model. From the graph, estimate the value of the boat after 5 years.

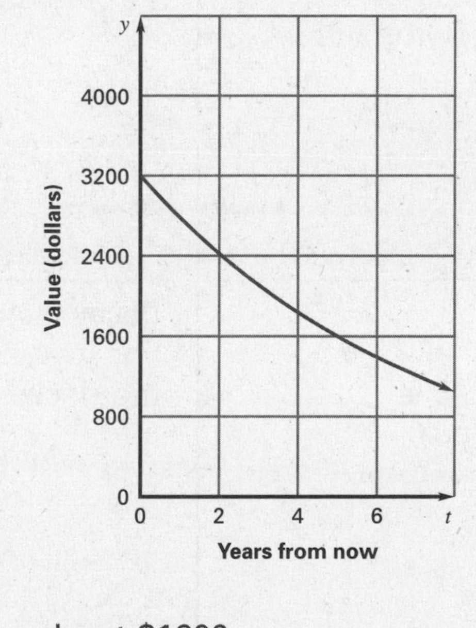

about \$1600

Words to Review

Give an example of the vocabulary word.

Exponential function $y = 2^x$	**Scientific notation** 4.5×10^{-7}
Exponential growth $y = 350(1.08)^t$	**Growth rate** For $y = 350(1.08)^t$, $1.08 - 1 = 0.08$ is the growth rate.
Growth factor For $y = 350(1.08)^t$, 1.08 is the growth factor.	**Exponential decay** $y = 475(0.85)^t$
Decay rate For $y = 475(0.85)^t$, $1 - 0.85 = 0.15$ is the decay rate.	**Decay factor** For $y = 475(0.85)^t$, 0.85 is the decay factor.

Review your notes and Chapter 8 by using the Chapter Review on pages 489–492 of your textbook.

Square Roots

Goal Evaluate and approximate square roots.

VOCABULARY

Square root If $b^2 = a$, then b is a square root of a.

Positive square root A positive square root, or principal square root, is the square root of a positive number that is itself positive.

Negative square root A negative square root is the negative number that is a square root of a positive number.

Radicand The number or expression inside a radical symbol is the radicand.

Perfect square The square of an integer is called a perfect square.

Radical expression A radical expression is an expression written with a radical symbol.

Example 1 *Find Square Roots of Numbers*

Evaluate the expression.

a. $-\sqrt{49} = -\underline{\sqrt{7^2}} = \underline{-7}$ — Negative square root

b. $\sqrt{49} = \underline{\sqrt{7^2}} = \underline{7}$ — Positive square root

c. $\sqrt{0} = \underline{0}$ — Square root of 0 is $\underline{0}$.

d. $\pm\sqrt{81} = \pm\underline{\sqrt{9^2}} = \underline{\pm 9}$ — Two square roots

The radical symbol is a grouping symbol. You must evaluate the expression inside the radical symbol before you find the square root.

Example 2 *Evaluate a Radical Expression*

Evaluate $\sqrt{b^2 - 4ac}$ when $a = -7$, $b = 8$, and $c = -1$.

$\sqrt{b^2 - 4ac} = \sqrt{\underline{8}^2 - 4(\underline{-7})(\underline{-1})}$ — Substitute for *a*, *b*, and *c*.

$= \sqrt{\underline{64} - \underline{28}}$ — Simplify.

$= \sqrt{\underline{36}}$ — Simplify.

$= \underline{6}$ — Positive square root

Checkpoint Evaluate the expression.

1. $\pm\sqrt{1}$	2. $\sqrt{121}$	3. $-\sqrt{9}$	4. $\sqrt{144}$
± 1	11	-3	12

Evaluate $\sqrt{b^2 - 4ac}$ for the given values.

5. $a = -8, b = 6, c = 2$	6. $a = 4, b = 9, c = 2$
10	7

9.2 Solving Quadratic Equations by Finding Square Roots

Goal Solve a quadratic equation by finding square roots.

VOCABULARY

Quadratic equation A quadratic equation is an equation that can be written in the standard form $ax^2 + bx + c = 0$, where $a \neq 0$.

SOLVING $x^2 = d$ BY FINDING SQUARE ROOTS

- If $d > 0$, then $x^2 = d$ has two solutions: $x = \pm\sqrt{d}$.
- If $d = 0$, then $x^2 = d$ has one solution: $x = 0$.
- If $d < 0$, then $x^2 = d$ has no real solution.

Example 1 ***Solve Quadratic Equations***

Remember that squaring a number and finding a square root of a number are inverse operations.

Solve the equation. Write the solutions as integers if possible. Otherwise, write them as radical expressions.

a. $k^2 = 17$

b. $p^2 = -4$

Solution

a. $k^2 = 17$ **Write original equation.**

$k = \pm\sqrt{17}$ **Find square roots.**

Answer The solutions are $\sqrt{17}$ and $-\sqrt{17}$.

b. $p^2 = -4$ has no real solution because the square of a real number is never negative.

Answer There is no real solution.

Example 2 *Rewrite Before Finding Square Roots*

Solve $4x^2 - 100 = 0$.

Solution

$4x^2 - 100 = 0$	**Write original equation.**
$4x^2 = \underline{100}$	**Add 100 to each side.**
$x^2 = \underline{25}$	**Divide each side by 4.**
$x = \underline{\pm\sqrt{25}}$	**Find square roots.**
$x = \underline{\pm 5}$	$\underline{5}^2 = 25$ **and** $(\underline{-5})^2 = 25$

Answer The solutions are $\underline{5}$ and $\underline{-5}$. Check both solutions in the original equation.

Checkpoint **Solve the equation or write *no real solution*. Write the solutions as integers if possible. Otherwise, write them as radical expressions.**

1. $x^2 = 13$ $\pm\sqrt{13}$	**2.** $x^2 = 16$ ± 4
3. $x^2 - 169 = 0$ ± 13	**4.** $54 - 6x^2 = 0$ ± 3

9.3 Simplifying Radicals

Goal Simplify radical expressions.

VOCABULARY

Simplest form of a radical expression The simplest form of a radical expression is an expression that has no perfect square factors other than 1 in the radicand, no fractions in the radicand, and no radicals in the denominator of a fraction.

PRODUCT PROPERTY OF RADICALS

$\sqrt{ab} = \underline{\sqrt{a}} \cdot \underline{\sqrt{b}}$ where $a \geq 0$ and $b \geq 0$

QUOTIENT PROPERTY OF RADICALS

$\sqrt{\frac{a}{b}} = \underline{\frac{\sqrt{a}}{\sqrt{b}}}$ where $a \geq 0$ and $b > 0$

Example 1 ***Simplify with the Product Property***

Simplify $\sqrt{80}$.

There can be more than one way to factor the radicand. An efficient method is to find the largest perfect square factor.

Solution

Look for perfect square factors to remove from the radicand.

$\sqrt{80} = \sqrt{\underline{16} \cdot 5}$	**Factor using perfect square factor.**
$= \underline{\sqrt{16}} \cdot \sqrt{5}$	**Use product property.**
$= \underline{4}\sqrt{5}$	**Simplify:** $\underline{\sqrt{16}} = \underline{4}$.

Example 2 *Simplify with the Quotient Property*

Simplify $\sqrt{\frac{20}{45}}$.

Solution

$\sqrt{\frac{20}{45}} = \sqrt{\frac{5 \cdot 4}{5 \cdot 9}}$ **Factor using perfect square factors.**

$= \sqrt{\frac{4}{9}}$ **Divide out common factors.**

$= \frac{\sqrt{4}}{\sqrt{9}}$ **Use quotient property.**

$= \frac{2}{3}$ **Simplify.**

Example 3 *Rationalize the Denominator*

Simplify $\sqrt{\frac{7}{48}}$.

Solution

$\sqrt{\frac{7}{48}} = \frac{\sqrt{7}}{\sqrt{48}}$ **Use quotient property.**

$= \frac{\sqrt{7}}{\boxed{\sqrt{16} \cdot \sqrt{3}}}$ **Use product property.**

$= \frac{\sqrt{7}}{\boxed{4\sqrt{3}}}$ **Remove perfect square factor.**

$= \frac{\sqrt{7}}{\boxed{4\sqrt{3}}} \cdot \frac{\sqrt{3}}{\sqrt{3}}$ **Multiply by a value of 1:** $\frac{\sqrt{3}}{\sqrt{3}}$ **= 1.**

$= \frac{\sqrt{21}}{12}$ **Simplify:** $4\sqrt{3} \cdot \sqrt{3} = 4 \cdot 3 = 12$.

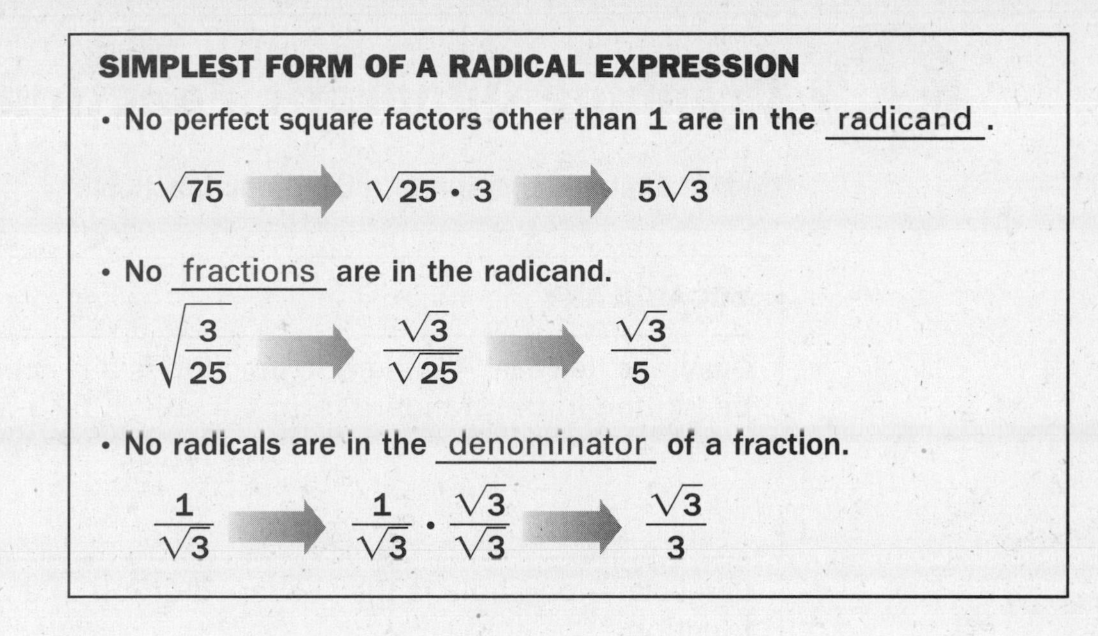

SIMPLEST FORM OF A RADICAL EXPRESSION

- No perfect square factors other than 1 are in the radicand.

$$\sqrt{75} \rightarrow \sqrt{25 \cdot 3} \rightarrow 5\sqrt{3}$$

- No fractions are in the radicand.

$$\sqrt{\frac{3}{25}} \rightarrow \frac{\sqrt{3}}{\sqrt{25}} \rightarrow \frac{\sqrt{3}}{5}$$

- No radicals are in the denominator of a fraction.

$$\frac{1}{\sqrt{3}} \rightarrow \frac{1}{\sqrt{3}} \cdot \frac{\sqrt{3}}{\sqrt{3}} \rightarrow \frac{\sqrt{3}}{3}$$

✓ *Checkpoint* **Simplify the expression.**

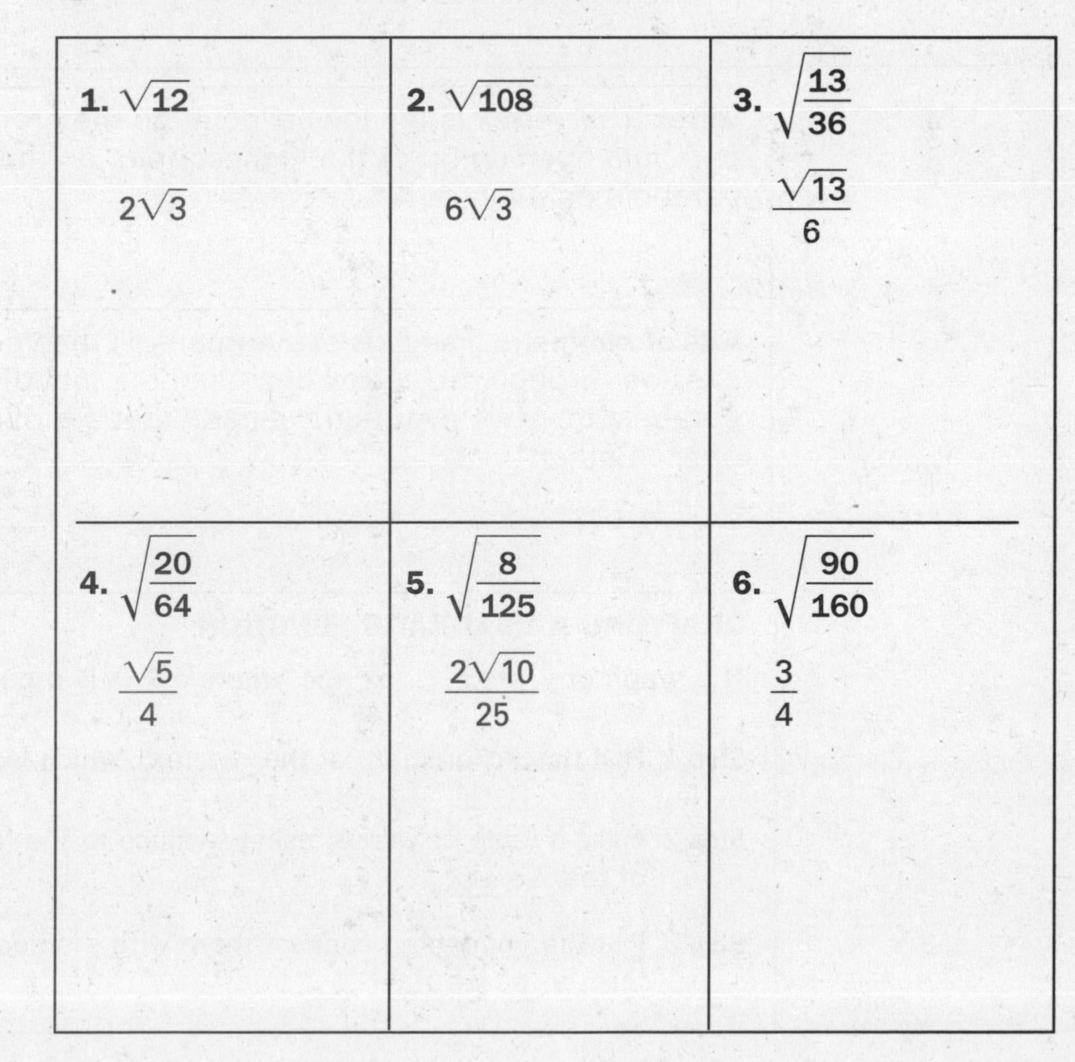

1. $\sqrt{12}$ $2\sqrt{3}$	**2.** $\sqrt{108}$ $6\sqrt{3}$	**3.** $\sqrt{\frac{13}{36}}$ $\frac{\sqrt{13}}{6}$
4. $\sqrt{\frac{20}{64}}$ $\frac{\sqrt{5}}{4}$	**5.** $\sqrt{\frac{8}{125}}$ $\frac{2\sqrt{10}}{25}$	**6.** $\sqrt{\frac{90}{160}}$ $\frac{3}{4}$

9.4 Graphing Quadratic Functions

Goal Sketch the graph of a quadratic function.

VOCABULARY

Quadratic function A quadratic function is a function that can be written in the standard form $y = ax^2 + bx + c$, where $a \neq 0$.

Parabola A parabola is the U-shaped graph of a quadratic function.

Vertex The vertex is the lowest point on the graph of a parabola opening up or the highest point on the graph of a parabola opening down.

Axis of symmetry The axis of symmetry is the vertical line passing through the vertex of a parabola that divides the parabola into two symmetrical parts that are mirror images of each other.

GRAPHING A QUADRATIC FUNCTION

The graph of $y = ax^2 + bx + c$, where $a \neq 0$, is a parabola.

Step 1 **Find the x-coordinate of the** vertex **, which is $x = -\frac{b}{2a}$.**

Step 2 **Make a table of values, using x-values to the left and right of the** vertex **.**

Step 3 **Plot the points and connect them with a smooth curve to form a** parabola **.**

Example 1 *Quadratic Function with Positive a-Value*

Sketch the graph of $y = x^2 - 4x + 4$.

In this quadratic equation, $a =$ <u>1</u>, $b =$ <u>-4</u>, and $c =$ <u>4</u>.

Find the x-coordinate of the vertex: $-\frac{b}{2a} = -\frac{-4}{2(1)} =$ <u>2</u>.

Make a table of values, using x-values to the left and right of $x =$ <u>2</u>.

x	-1	0	1	2	3	4	5
y	9	4	1	0	1	4	9

The axis of symmetry passes through the vertex. So the axis of symmetry of this parabola is the vertical line $x = 2$.

Plot the points. The vertex is (<u>2</u>, <u>0</u>). Connect the points to form a parabola that opens <u>up</u> because a is <u>positive</u>.

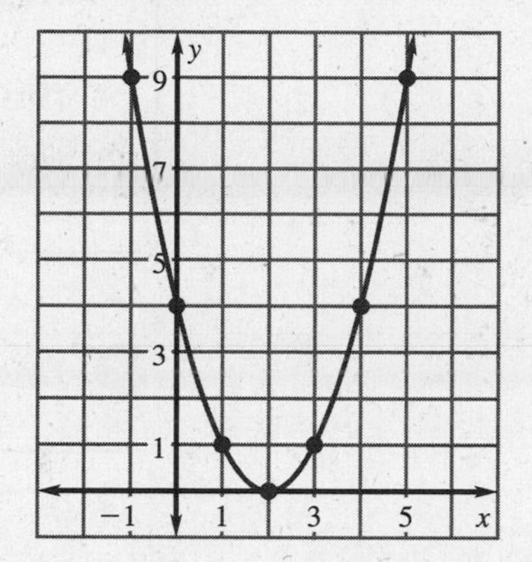

Example 2 *Quadratic Function with Negative a-Value*

Sketch the graph of $y = -x^2 + 2x - 1$.

In this quadratic equation, $a =$ <u>-1</u>, $b =$ <u>2</u>, and $c =$ <u>-1</u>.

Find the x-coordinate of the vertex: $-\frac{b}{2a} = -\frac{2}{2(-1)} =$ <u>1</u>.

Make a table of values, using x-values to the left and right of $x =$ <u>1</u>.

x	-2	-1	0	1	2	3	4
y	-9	-4	-1	0	-1	-4	-9

Plot the points. The vertex is (<u>1</u>, <u>0</u>). Connect the points to form a parabola that opens <u>down</u> because a is <u>negative</u>.

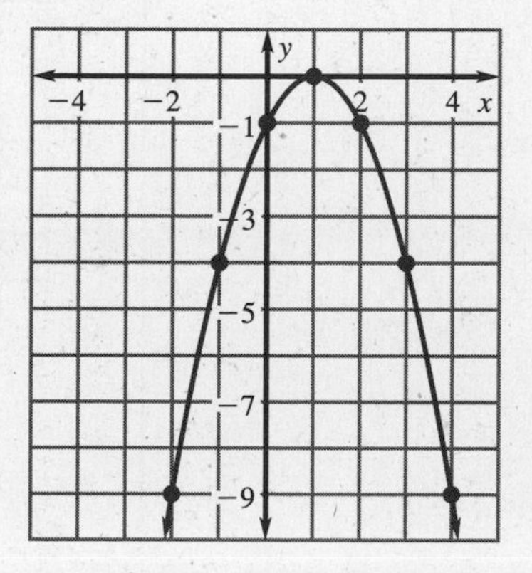

GRAPH OF A QUADRATIC FUNCTION

- The graph of $y = ax^2 + bx + c$ is a parabola.
- If a is positive, the parabola opens up.
- If a is negative, the parabola opens down.
- The vertex has an x-coordinate of $-\frac{b}{2a}$.
- The axis of symmetry is the vertical line $x = -\frac{b}{2a}$.
- The y-intercept is c.

✔ *Checkpoint* **Decide whether the parabola opens *up* or *down*.**

1. $y = -x^2 - x + 1$	**2.** $y = 3x^2 - 2x + 2$
down	up

Sketch the graph of the function.

3. $y = x^2 + 2x - 5$

4. $y = -4x^2 + 4x - 1$

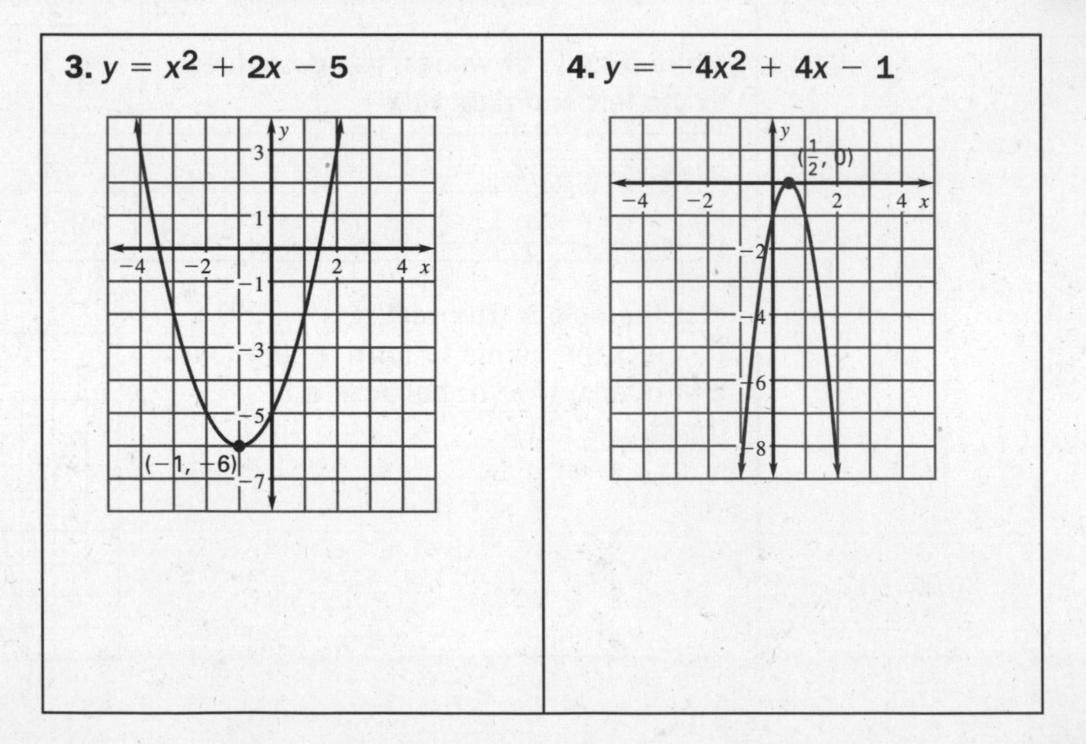

9.5 Solving Quadratic Equations by Graphing

Goal Use a graph to find or check a solution of a quadratic equation.

VOCABULARY

Roots of a quadratic equation The solutions, or roots, of a quadratic equation are the x-intercepts of the graph.

ESTIMATING SOLUTIONS BY GRAPHING

The solutions of a quadratic equation in one variable x can be estimated by graphing. Use the following steps:

Step 1 **Write the equation in the form** $ax^2 + bx + c = 0$.

Step 2 **Sketch the graph of the related quadratic function** $y = ax^2 + bx + c$.

Step 3 **Estimate the values of the** x-intercepts**, if any.**

Example 1 ***Use a Graph to Solve an Equation***

Use the graph of $y = \frac{1}{4}x^2 - 9$ to estimate the solutions of $\frac{1}{4}x^2 - 9 = 0$.

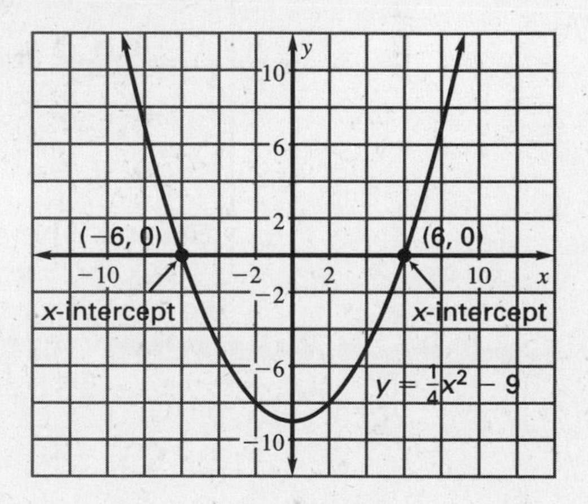

The graph appears to intersect the x-axis at (−6 , 0) and (6 , 0). By substituting $x =$ −6 and $x =$ 6 in $\frac{1}{4}x^2 - 9 = 0$, you can check that −6 and 6 are solutions of the given equation.

Example 2 *Solve an Equation by Graphing*

Use a graph to estimate the solutions of $x^2 + 2x = 3$. Check your solutions algebraically.

Solution

1. Write the equation in the standard form $ax^2 + bx + c = 0$.

$x^2 + 2x = 3$ **Write original equation.**

$x^2 + 2x \underline{- 3} = \underline{0}$ **Subtract $\underline{3}$ from each side.**

2. Sketch the graph of the related quadratic function $y = \underline{x^2 + 2x - 3}$.

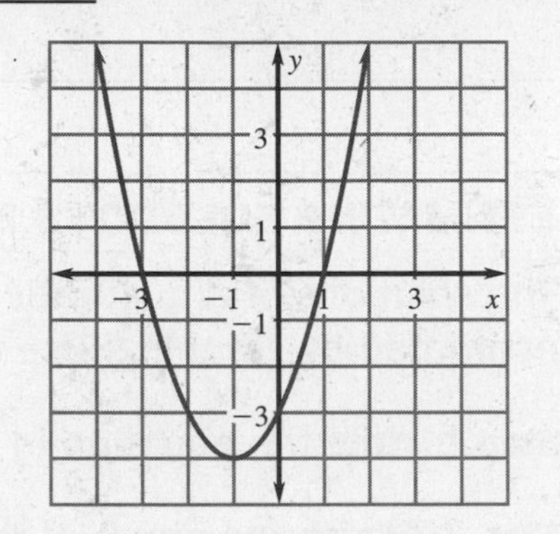

3. Estimate the values of the x-intercepts. From the graph, the x-intercepts appear to be $x = \underline{-3}$ and $x = \underline{1}$.

Check You can check your solutions algebraically using substitution.

Check $x = \underline{-3}$:

$x^2 + 2x = 3$

$(\underline{-3})^2 + 2(\underline{-3}) \stackrel{?}{=} 3$

$\underline{9 - 6} \stackrel{?}{=} 3$

$\underline{3} = 3$

Check $x = \underline{1}$:

$x^2 + 2x = 3$

$(\underline{1})^2 + 2(\underline{1}) \stackrel{?}{=} 3$

$\underline{1 + 2} \stackrel{?}{=} 3$

$\underline{3} = 3$

Answer The solutions are $\underline{-3}$ and $\underline{1}$.

Checkpoint Complete the following exercise.

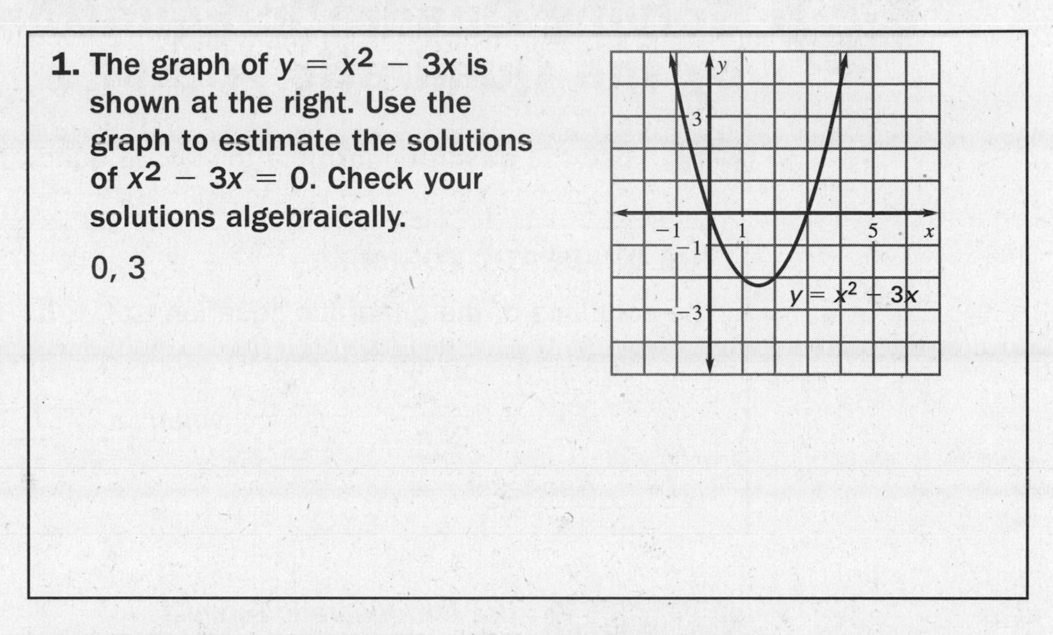

1. The graph of $y = x^2 - 3x$ is shown at the right. Use the graph to estimate the solutions of $x^2 - 3x = 0$. Check your solutions algebraically.

   0, 3

Use a graph to estimate the solutions of the equation. Check your solutions algebraically.

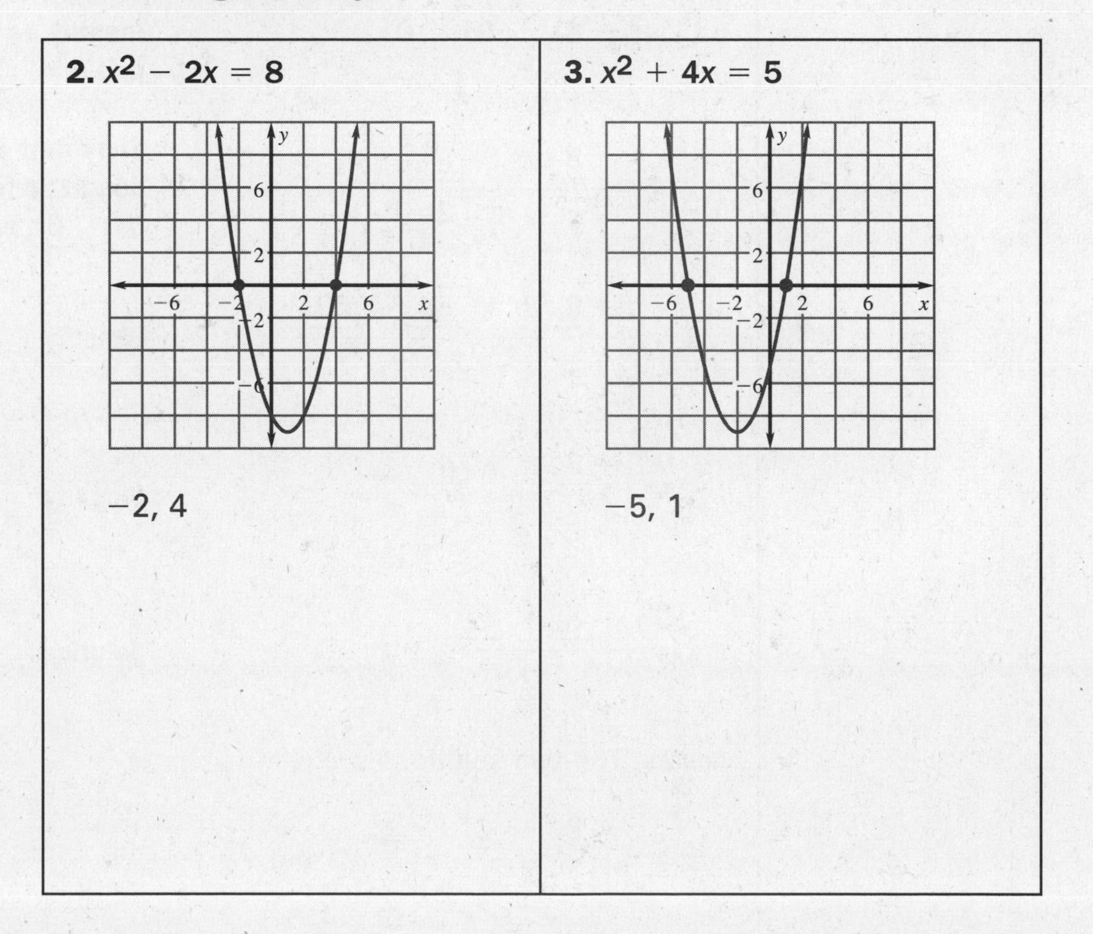

2. $x^2 - 2x = 8$

   −2, 4

3. $x^2 + 4x = 5$

   −5, 1

9.6 Solving Quadratic Equations by the Quadratic Formula

Goal Use the quadratic formula to solve a quadratic equation.

THE QUADRATIC FORMULA

The solutions of the quadratic equation $ax^2 + bx + c = 0$ are

$$x = \frac{-\underline{b} + \sqrt{\underline{b^2} - 4\,\underline{ac}}}{2\,\underline{a}}$$ when $\underline{a} \neq \underline{0}$ and

$\underline{b^2 - 4ac} \geq \underline{0}$.

Example 1 *Use the Quadratic Formula*

Solve $x^2 + 8x + 15 = 0$ by using the quadratic formula.

$(1)x^2 + 8x + 15 = 0$ — **Identify $a = \underline{1}$, $b = \underline{8}$, and $c = \underline{15}$.**

$$x = \frac{-\underline{8} \pm \sqrt{\underline{8^2} - 4(\underline{1})(\underline{15})}}{2(\underline{1})}$$

Substitute values in the quadratic formula: $a = \underline{1}$, $b = \underline{8}$, and $c = \underline{15}$.

$$x = \frac{-\underline{8} \pm \sqrt{\underline{64} - \underline{60}}}{\underline{2}}$$

Simplify.

$$x = \frac{-\underline{8} \pm \sqrt{\underline{4}}}{\underline{2}}$$

Simplify.

$$x = \frac{-\underline{8} \pm \underline{2}}{\underline{2}}$$

Solutions

Answer The two solutions are

$$x = \frac{-\underline{8} + \underline{2}}{\underline{2}} = \underline{-3} \text{ and } x = \frac{-\underline{8} - \underline{2}}{\underline{2}} = \underline{-5}.$$

Example 2 ***Write in Standard Form***

Solve $3x^2 - 7x = 11$. Round the results to the nearest hundredth.

Solution

$3x^2 - 7x = 11$	**Write original equation.**
$3x^2 - 7x - 11 = 0$	**Rewrite equation in standard form.**
$x = \frac{-(-7) \pm \sqrt{(-7)^2 - 4(3)(-11)}}{2(3)}$	**Substitute values in the quadratic formula.**
$x = \frac{7 \pm \sqrt{49 + 132}}{6}$	**Simplify.**
$x = \frac{7 \pm \sqrt{181}}{6}$	**Solutions**

Answer The equation has two solutions:

$x = \frac{7 + \sqrt{181}}{6} \approx$ 3.41 and $x = \frac{7 - \sqrt{181}}{6} \approx$ −1.08 .

✔ ***Checkpoint*** **Use the quadratic formula to solve the equation. If the solution involves radicals, round to the nearest hundredth.**

1. $x^2 + 4x - 5 = 0$	**2.** $3x^2 - 8x = 9$
−5, 1	$\frac{8 + \sqrt{172}}{6} \approx 3.52$, $\frac{8 - \sqrt{172}}{6} \approx -0.85$

Example 3 *Model Vertical Motion*

> Velocity indicates speed and direction (up is positive and down is negative). Speed is the absolute value of velocity.

Diving A cliff diver jumps from a height of 58 feet above the water with an initial velocity of 5 feet per second. How long will it take the diver to reach the water?

Solution

The diver's initial velocity is $v =$ 5 feet per second and the diver's initial height is $s =$ 58 feet . The diver will reach the water when the height is 0 feet .

$h = -16t^2 + vt + s$	**Choose a vertical motion model.**
$h = -16t^2 + 5t + 58$	**Substitute values.**
$0 = -16t^2 + 5t + 58$	**Substitute 0 for *h*. Write in standard form.**
$t = \frac{-b \pm \sqrt{b^2 - 4ac}}{2a}$	**Write quadratic formula.**
$t = \frac{-5 \pm \sqrt{5^2 - 4(-16)(58)}}{2(-16)}$	**Substitute for *a*, *b*, and *c*.**
$t = \frac{-5 \pm \sqrt{3737}}{-32}$	**Simplify.**
$t \approx$ 2.07 or -1.75	**Evaluate the radical expressions.**

Answer Because time cannot be a negative number, disregard the solution -1.75 . So, the diver will reach the water in 2.07 seconds.

✓ ***Checkpoint*** **Complete the following exercise.**

3. A tennis ball is dropped from the top of a building, which is 40 feet above the ground. How long will it take the tennis ball to reach the ground?

$\frac{\sqrt{10}}{2} \approx 1.58$ sec

9.7 Using the Discriminant

Goal Use the discriminant to determine the number of solutions of a quadratic equation.

VOCABULARY

Discriminant The discriminant is the expression $b^2 - 4ac$, where a, b, and c are the coefficients of the quadratic equation $ax^2 + bx + c = 0$.

Recall that positive real numbers have two square roots, zero has only one square root, and negative numbers have no real square roots.

THE NUMBER OF SOLUTIONS OF A QUADRATIC EQUATION

Consider the quadratic equation $ax^2 + bx + c = 0$.

- If the value of $b^2 - 4ac$ is positive, then the equation has two solutions.
- If the value of $b^2 - 4ac$ is zero, then the equation has one solution.
- If the value of $b^2 - 4ac$ is negative, then the equation has no real solution.

Example 1 ***Find the Number of Solutions***

Determine the number of solutions of $x^2 - 2x - 9 = 0$ by finding the value of the discriminant.

Solution

Use the standard form of a quadratic equation, $ax^2 + bx + c = 0$, to identify the coefficients.

$x^2 - 2x - 9 = 0$ **Identify $a =$ 1, $b =$ −2, and $c =$ −9.**

$b^2 - 4ac = (-2)^2 - 4(1)(-9)$

$= 4 + 36$

$= 40$

Answer The discriminant is positive, so the equation has two solutions.

Example 2 *Find the Number of Solutions*

Find the value of the discriminant. Then use the value to determine whether the equation has *two solutions*, *one solution*, or *no real solution*.

a. $x^2 - 8x + 16 = 0$

b. $-3x^2 + 4x - 5 = 0$

Solution

a. $x^2 - 8x + 16 = 0$ — **Identify** $a =$ 1, $b =$ −8, **and** $c =$ 16.

$b^2 - 4ac = (-8)^2 - 4(1)(16)$ — **Substitute.**

$= 64 - 64$ — **Simplify.**

$= 0$ — **Discriminant is** zero.

Answer The discriminant is zero, so the equation has one solution.

b. $-3x^2 + 4x - 5 = 0$ — **Identify** $a =$ −3, $b =$ 4, **and** $c =$ −5.

$b^2 - 4ac = (4)^2 - 4(-3)(-5)$ — **Substitute.**

$= 16 - 60$ — **Simplify.**

$= -44$ — **Discriminant is** negative.

Answer The discriminant is negative, so the equation has no real solution.

✓ *Checkpoint* **Determine whether the equation has *two solutions*, *one solution*, or *no real solution*.**

1. $-x^2 - 5x - 9 = 0$ no real solution	**2.** $4x^2 - 4x + 1 = 0$ one solution
3. $8x^2 + 8x + 1 = 0$ two solutions	**4.** $\frac{1}{2}x^2 - x - 4 = 0$ two solutions

Example 3 *Find the Number of x-Intercepts*

Determine whether the graph of the function will intersect the x-axis in *zero*, *one*, or *two* points.

a. $y = x^2 - 2x + 1$ **b.** $y = x^2 - 2x + 3$

Solution

a. Let $y = 0$. Then find the value of the discriminant.

$0 = x^2 - 2x + 1$ **Identify** a **=** 1, b **=** −2, **and** c **=** 1.

$b^2 - 4ac = (-2)^2 - 4(1)(1)$ **Substitute.**

$= 4 - 4$ **Simplify.**

$= 0$ **Discriminant is** zero.

Answer The discriminant is zero, so the equation has one solution *and* the graph will intersect the x-axis in one point.

b. Let $y = 0$. Then find the value of the discriminant.

$0 = x^2 - 2x + 3$ **Identify** a **=** 1, b **=** −2, **and** c **=** 3.

$b^2 - 4ac = (-2)^2 - 4(1)(3)$ **Substitute.**

$= 4 - 12$ **Simplify.**

$= -8$ **Discriminant is** negative.

Answer The discriminant is negative, so the equation has no real solution *and* the graph will intersect the x-axis in zero points.

✓ ***Checkpoint*** **Determine whether the graph of the function will intersect the x-axis in *zero*, *one*, or *two* points.**

5. $y = x^2 - 2x - 1$	**6.** $y = x^2 - 4x + 5$
two	zero

9.8 Graphing Quadratic Inequalities

Goal Sketch the graph of a quadratic inequality in two variables.

VOCABULARY

Quadratic inequalities A quadratic inequality is an inequality that can be written in one of the following forms.

$y < ax^2 + bx + c$ $\quad y < ax^2 + bx + c$

$y > ax^2 + bx + c$ $\quad y \geq ax^2 + bx + c$

Graph of a quadratic inequality The graph of a quadratic inequality consists of the graph of all ordered pairs (x, y) that are solutions of the inequality.

Example 1 ***Check Points***

Sketch the graph of $y = x^2 + 2x - 3$. Plot and label the points $A(0, 1)$, $B(-2, 3)$, and $C(1, -3)$. Determine whether each point lies inside or outside the parabola.

Solution

1. **Sketch** the graph of $y = x^2 + 2x - 3$.
2. **Plot** and label the points $A(0, 1)$, $B(-2, 3)$, and $C(1, -3)$.

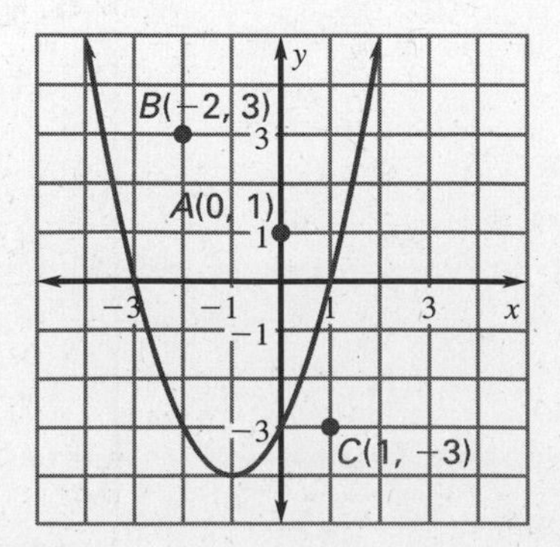

Answer Points <u>A</u> and <u>B</u> lie inside the parabola. Point <u>C</u> lies outside.

METHOD 1: GRAPHING A QUADRATIC INEQUALITY

Step 1 Sketch the graph of $y = ax^2 + bx + c$ that corresponds to the inequality.

Sketch a dashed parabola for inequalities with $<$ or $>$ to show that the points on the parabola are not solutions.

Sketch a solid parabola for inequalities with $\leq$ or $\geq$ to show that the points on the parabola are solutions.

Step 2 The parabola separates the coordinate plane into two regions. Test a point that is *not* on the parabola to determine whether the point is a solution of the inequality.

Step 3 If the test point is a solution, shade its region. If not, shade the other region.

Example 2 ***Graph a Quadratic Inequality***

Sketch the graph of $y < x^2 - 4x$.

1. Sketch the equation $y = x^2 - 4x$ that corresponds to the inequality $y < x^2 - 4x$. Use a dashed line since the inequality contains the symbol $<$.

If the point (0, 0) is not on the parabola, then (0, 0) is usually good to use as a test point.

2. Test a point such as (0, 1) that is *not* on the parabola.

$y < x^2 - 4x$	**Write original inequality.**
$1 \overset{?}{<} 0^2 - 4(0)$	**Substitute 0 for *x* and 1 for *y*.**
$1 \not< 0$	**1 is not less than 0.**

Because 1 is not less than 0, the ordered pair (0, 1) is not a solution.

3. Shade the region outside the parabola. The point (0, 1) is inside the parabola and it is not a solution, so the graph of $y < x^2 - 4x$ is all points that are outside, but not on, the parabola.

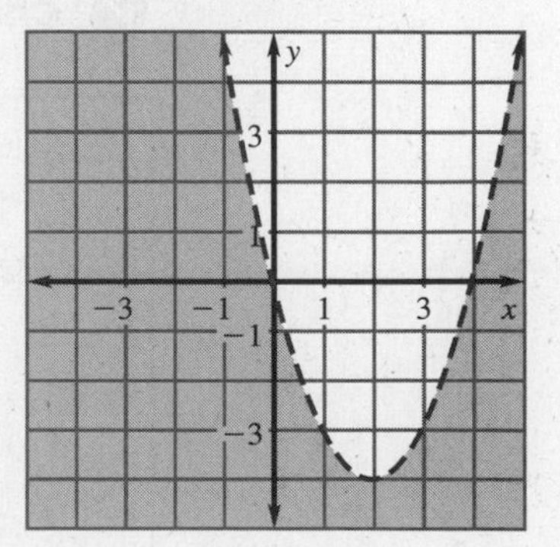

METHOD 2: GRAPHING A QUADRATIC INEQUALITY

Step 1 Sketch the graph of $y = ax^2 + bx + c$, using a dashed or a solid curve as in the previous method.

Step 2 If the inequality is $y > ax^2 + bx + c$ or $y \geq ax^2 + bx + c$, shade the region above the parabola.

If the inequality is $y < ax^2 + bx + c$ or $y \leq ax^2 + bx + c$, shade the region below the parabola.

Example 3 *Graph a Quadratic Inequality*

Sketch the graph of $y \leq -x^2 - 2x + 3$.

1. **Sketch** the graph of the equation $y = -x^2 - 2x + 3$. The x-coordinate of the vertex is $-\frac{b}{2a}$, or -1. Make a table of values, using x-values to the left and right of $x = -1$.

x	−4	−3	−2	−1	0	1	2
y	−5	0	3	4	3	0	−5

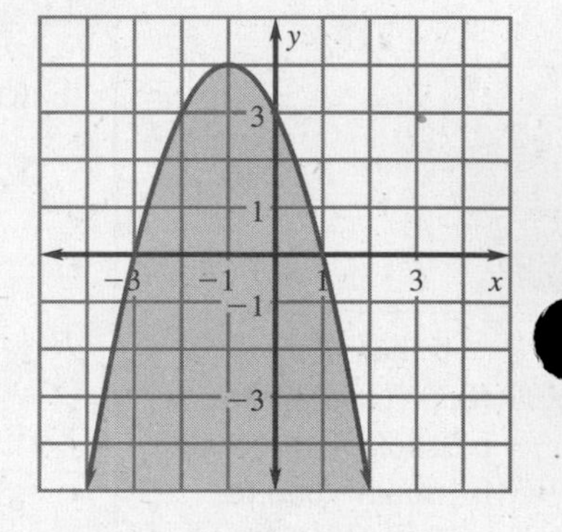

Plot the points and connect them with a smooth curve to form a parabola. Use a solid line since the inequality contains $\leq$.

2. **Shade** the region below the parabola because the inequality states that y is less than or equal to $-x^2 - 2x + 3$.

Checkpoint Sketch the graph of the inequality.

1. $y > -x^2 + 4$

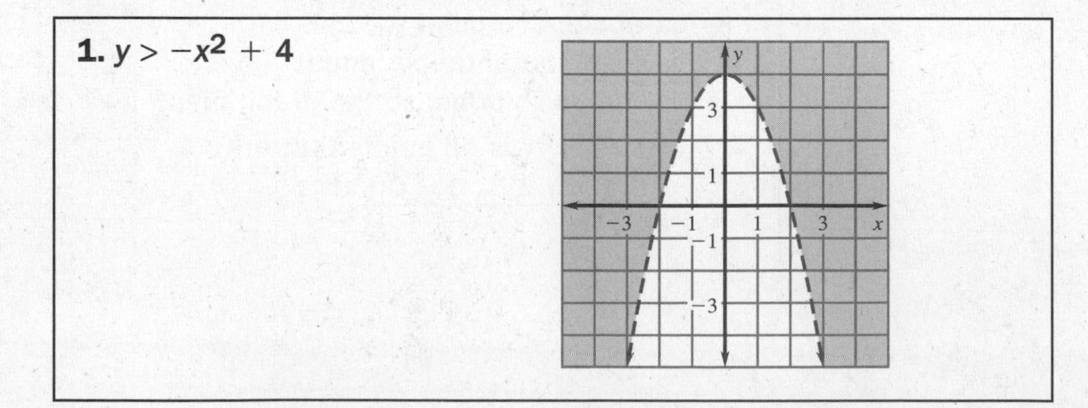

Words to Review

Give an example of the vocabulary word.

Square root	**Positive and negative square roots**
2 is a square root of 4.	positive: $\sqrt{4} = 2$ negative: $-\sqrt{4} = -2$
Radicand For $\sqrt{4}$, the radicand is 4.	**Perfect square** 36
Radical expression $\sqrt{a + b}$	**Quadratic equation in standard form** $2x^2 - x - 3 = 0$
Simplest form of a radical expression $2\sqrt{5}$ is in simplest form.	**Quadratic function in standard form** $y = 2x^2 - x - 3$

Parabola 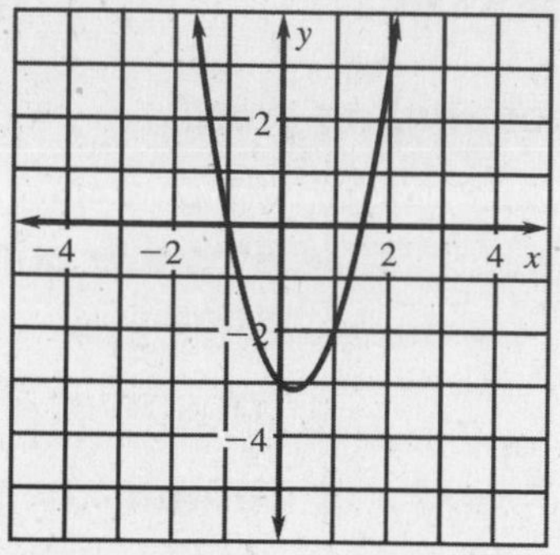	**Vertex** For the parabola $y = x^2$, the vertex is (0, 0).
Axis of symmetry For the parabola $y = x^2$, the axis of symmetry is $x = 0$.	**Roots of a quadratic equation** For $y = 2x^2 - x - 3$, the roots are -1 and $\frac{3}{2}$.
Quadratic formula The solutions of $2x^2 - x - 3 = 0$, using the quadratic formula, are $\frac{-(-1) \pm \sqrt{(-1)^2 - 4(2)(-3)}}{2(2)}$.	**Discriminant** For $y = 2x^2 - x - 3$, the discriminant is $(-1)^2 - 4(2)(-3)$.
Quadratic inequality $y \geq 2x^2 - x - 3$	**Graph of a quadratic inequality** 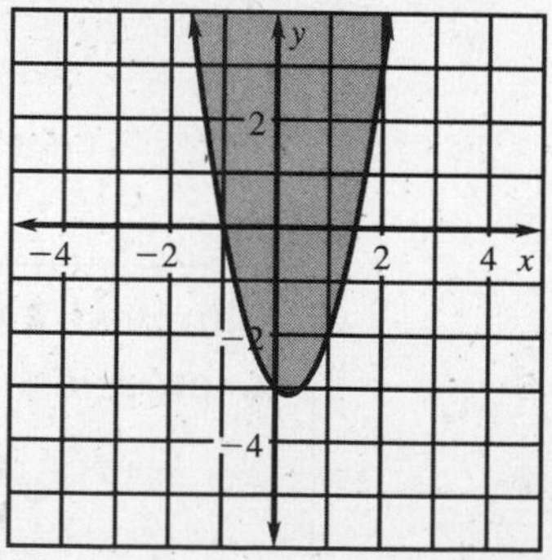

Review your notes and Chapter 9 by using the Chapter Review on pages 553–556 of your textbook.

10.1 Adding and Subtracting Polynomials

Goal Add and subtract polynomials.

VOCABULARY

Monomial A monomial is a number, a variable, or a product of a number and one or more variables with whole number exponents. A monomial is a polynomial with only one term.

Degree of a monomial The degree of a monomial is the exponent of the variable in the monomial. The degree of $5x^2$ is 2.

Polynomial A polynomial is a monomial or a sum of monomials.

Binomial A binomial is a polynomial consisting of two terms.

Trinomial A trinomial is a polynomial consisting of three terms.

Standard form of a polynomial in one variable The standard form of a polynomial in one variable is a polynomial whose terms are written in decreasing order, from largest exponent to smallest exponent.

Degree of a polynomial in one variable The degree of a polynomial in one variable is the largest exponent of that variable.

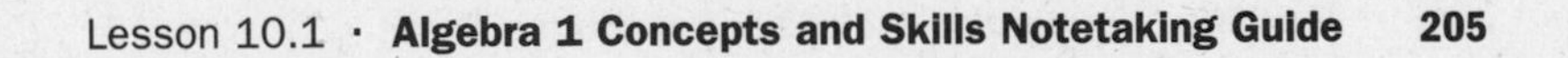

Example 1 *Find the Degree of a Monomial*

a. $-2x^6$ — The exponent of x is 6, so the degree is 6.

b. $\frac{3}{7}y$ — The exponent of y is 1, so the degree is 1.

c. 23 — The exponent is 0, so the degree is 0.

Example 2 *Identify Polynomials*

Polynomial	Degree	Identified by Degree	Identified by Number of Terms
a. -3	0	constant	monomial
b. $-x + 1$	1	linear	binomial
c. $x^2 + 3$	2	quadratic	binomial
d. $5x^3 - 3x^2 + x - 8$	3	cubic	polynomial

Example 3 *Add Polynomials*

To add two polynomials, add the like terms. You can use a vertical format or a horizontal format.

Find the sum. Write the answer in standard form.

a. $(-3x^3 + 11x^2 - 8x + x^5 + 2) + (8x - 2x^4 + 7x^3 - 3 + 12x^2)$

b. $(3x^2 - 2x + 3) + (x + 4x^2 - 1)$

Solution

a. **Vertical format:** Write each expression in standard form. Line up like terms vertically.

$$\begin{array}{r} x^5 \qquad\quad - 3x^3 + 11x^2 - 8x + 2 \\ -2x^4 + 7x^3 + 12x^2 + 8x - 3 \\ \hline \boxed{x^5 - 2x^4 + 4x^3 + 23x^2 \qquad - 1} \end{array}$$

b. **Horizontal format:** Group like terms.

$(3x^2 - 2x + 3) + (x + 4x^2 - 1)$

$= (3x^2 + 4x^2) + (-2x + x) + (3 - 1)$

$= 7x^2 - x + 2$

Example 4 *Subtract Polynomials*

Find the difference. Write the answer in standard form.

$(11x^4 + x^3 - x + 5) - (-x^4 - x^2 + 2x + 8)$

Solution

Use a vertical format. To subtract one polynomial from another, you add the opposite. One way to do this is to multiply each term in the subtraction polynomial by -1 and line up the like terms vertically. Then add.

$$\begin{array}{r} (11x^4 + x^3 \quad\quad - x + 5) \\ -(-x^4 - x^2 + 2x + 8) \\ \hline \end{array} \quad \textbf{Add the opposite.} \quad \begin{array}{r} 11x^4 + x^3 \quad\quad\quad - x + 5 \\ \boxed{+}\ x^4 \quad\quad \boxed{+}\,x^2 \boxed{-}\,2x \boxed{-}\,8 \\ \hline \boxed{12x^4 + x^3 + x^2 - 3x - 3} \end{array}$$

Checkpoint Find the sum or difference.

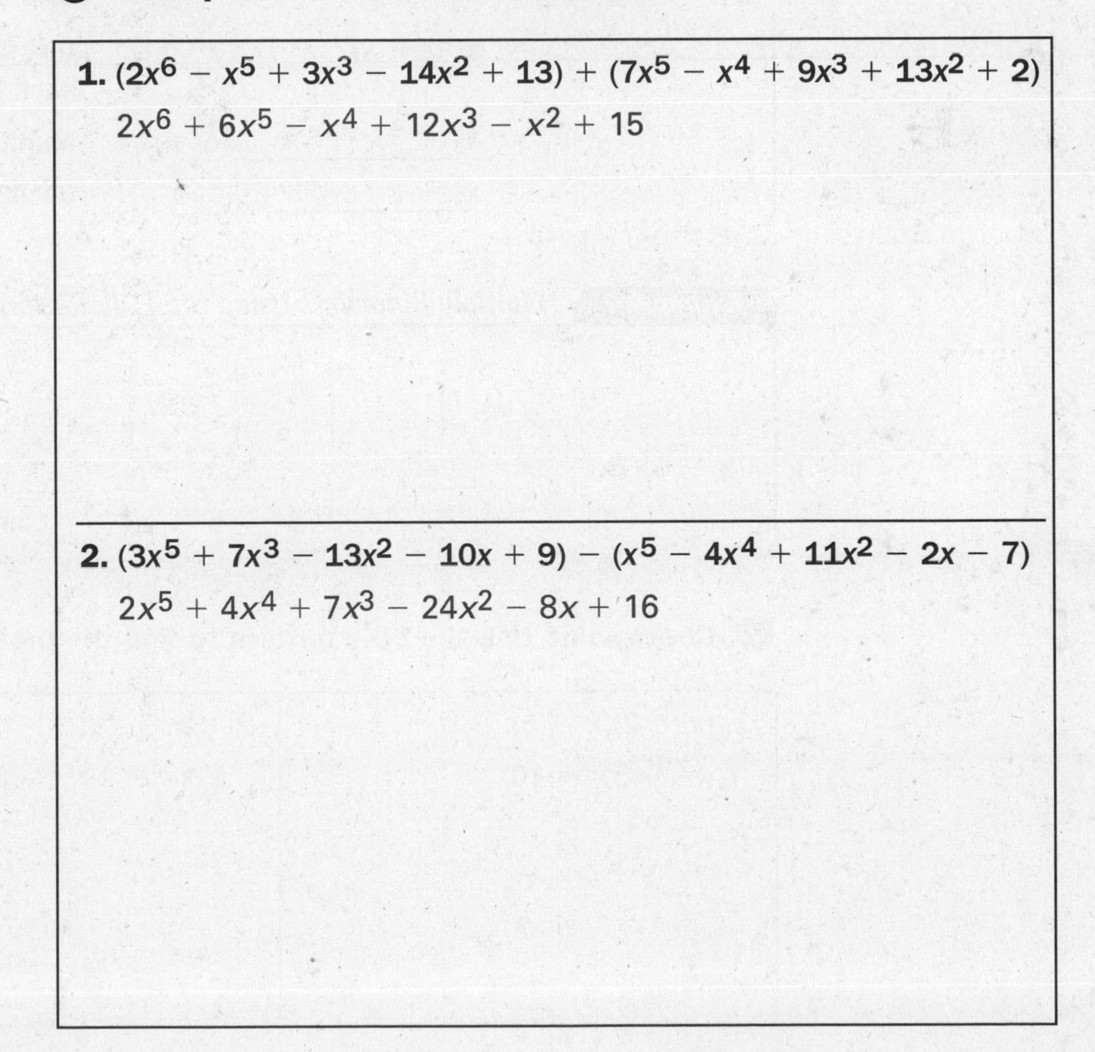

1. $(2x^6 - x^5 + 3x^3 - 14x^2 + 13) + (7x^5 - x^4 + 9x^3 + 13x^2 + 2)$

$2x^6 + 6x^5 - x^4 + 12x^3 - x^2 + 15$

2. $(3x^5 + 7x^3 - 13x^2 - 10x + 9) - (x^5 - 4x^4 + 11x^2 - 2x - 7)$

$2x^5 + 4x^4 + 7x^3 - 24x^2 - 8x + 16$

10.2 Multiplying Polynomials

Goal Multiply polynomials.

VOCABULARY

FOIL pattern The FOIL pattern is a pattern used to multiply two binomials. Multiply the First, Outer, Inner, and Last terms.

Example 1 ***Use the Distributive Property***

$(x - 5)(x + 7) = \underline{x}(x + 7) - \underline{5}(x + 7)$ **Distribute $(x + 7)$ to each term of $(x - 5)$.**

$= \underline{x(x) + x(7) - 5(x) - 5(7)}$ **Distribute x and -5 to each term of $(x + 7)$.**

$= \underline{x^2 + 7x - 5x - 35}$ **Multiply.**

$= \underline{x^2 + 2x - 35}$ **Combine like terms.**

Example 2 ***Multiply Binomials Using the FOIL Pattern***

F O I L

$(x - 5)(7x + 1) = \underline{7x^2} + \underline{x} - \underline{35x} - \underline{5}$

$= \underline{7x^2} - \underline{34x} - \underline{5}$ **Combine like terms.**

Checkpoint **Use the FOIL pattern to find the product.**

1. $(x + 2)(x - 5)$	**2.** $(3x - 1)(2x - 3)$
$x^2 - 3x - 10$	$6x^2 - 11x + 3$

Example 3 *Multiply Polynomials Vertically*

Find the product $(4 - x)(8 - 11x + x^2)$.

Solution

Line up like terms vertically. Then multiply.

	$x^2 - 11x + 8$	**Standard form**
$\times$	$-x + 4$	**Standard form**
	$4x^2 - 44x + 32$	
	$-x^3 + 11x^2 - 8x$	
	$-x^3 + 15x^2 - 52x + 32$	

To multiply two polynomials, remember that each term of one polynomial must be multiplied by each term of the other polynomial.

Example 4 *Multiplying Polynomials Horizontally*

Find the product $(2x^3 - 9x^2 - 11x)(2 - x^2)$.

Solution

Multiply $2 - x^2$ by each term of $2x^3 - 9x^2 - 11x$.

$(2x^3 - 9x^2 - 11x)(2 - x^2)$

$= 2x^3(2 - x^2) - 9x^2(2 - x^2) - 11x(2 - x^2)$ **Use distributive property.**

$= 4x^3 - 2x^5 - 18x^2 + 9x^4 - 22x + 11x^3$ **Use distributive property.**

$= -2x^5 + 9x^4 + (4x^3 + 11x^3) - 18x^2 - 22x$ **Group like terms.**

$= -2x^5 + 9x^4 + 15x^3 - 18x^2 - 22x$ **Combine like terms.**

Remember to use the product of powers property when multiplying two variable terms.

✔ *Checkpoint* **Use a vertical format to find the product. Write your answer in standard form.**

3. $(x + 5)(x^2 - 2x + 3)$ $x^3 + 3x^2 - 7x + 15$	**4.** $(x - 2)(2x^2 - 6 - x)$ $2x^3 - 5x^2 - 4x + 12$

Use a horizontal format to find the product. Write your answer in standard form.

5. $(x - 4)(x^2 - 7x + 4)$ $x^3 - 11x^2 + 32x - 16$	**6.** $(11x^2 + 7x - 3)(-5x + 1)$ $-55x^3 - 24x^2 + 22x - 3$

10.3 Special Products of Polynomials

Goal Use special product patterns to multiply polynomials.

SPECIAL PRODUCT PATTERNS

Sum and Difference Pattern

$(a + b)(a - b) = \underline{a^2 - b^2}$

Example: $(3x - 4)(3x + 4) = \underline{9x^2 - 16}$

Square of a Binomial Pattern

$(a + b)^2 = \underline{a^2 + 2ab + b^2}$

Example: $(x + 4)^2 = \underline{x^2 + 8x + 16}$

$(a - b)^2 = \underline{a^2 - 2ab + b^2}$

Example: $(2x - 6)^2 = \underline{4x^2 - 24x + 36}$

Example 1 ***Use the Sum and Difference Pattern***

Find the product $(9w + 3)(9w - 3)$.

$(a + b)(a - b) = \underline{a^2 - b^2}$ **Write pattern.**

$(9w + 3)(9w - 3) = \underline{(9w)^2 - 3^2}$ **Apply pattern.**

$= \underline{81w^2 - 9}$ **Simplify.**

You can use the FOIL pattern to check your answer.

Example 2 ***Use the Square of a Binomial Pattern***

Find the product.

a. $(12x + 4)^2$

b. $(3k - 2m)^2$

Solution

a. $(a + b)^2 = \underline{a^2 + 2ab + b^2}$ **Write pattern.**

$(12x + 4)^2 = \underline{(12x)^2 + 2(12x)(4) + 4^2}$ **Apply pattern.**

$= \underline{144x^2 + 96x + 16}$ **Simplify.**

b. $(a - b)^2 = \underline{a^2 - 2ab + b^2}$ **Write pattern.**

$(3k - 2m)^2 = \underline{(3k)^2 - 2(3k)(2m) + (2m)^2}$ **Apply pattern.**

$= \underline{9k^2 - 12km + 4m^2}$ **Simplify.**

Checkpoint Find the product.

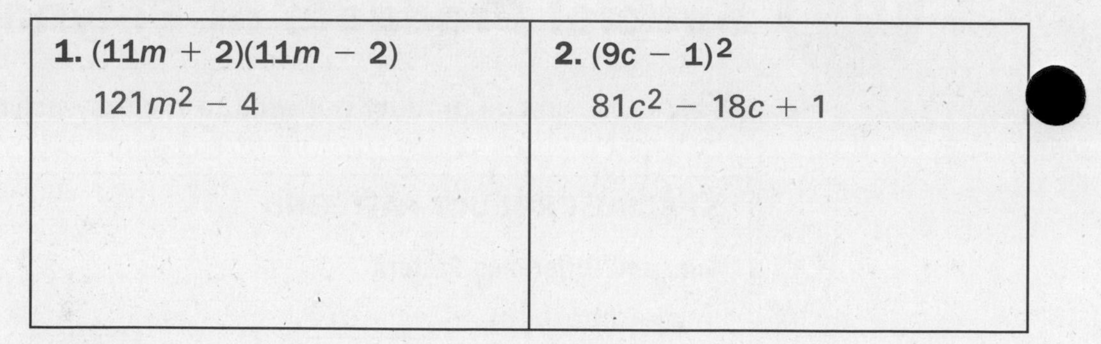

1. $(11m + 2)(11m - 2)$	2. $(9c - 1)^2$
$121m^2 - 4$	$81c^2 - 18c + 1$

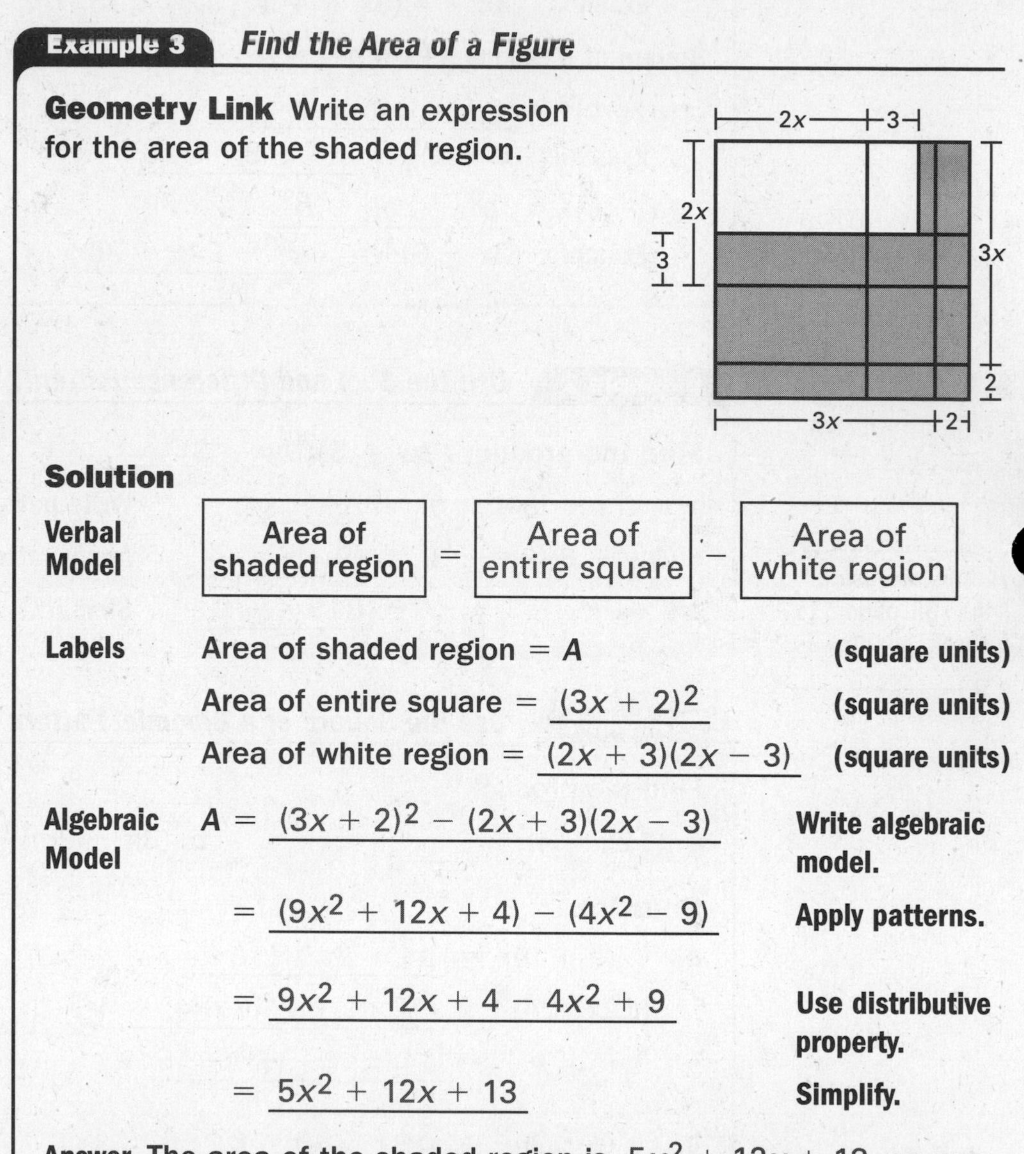

Example 3 ***Find the Area of a Figure***

Geometry Link Write an expression for the area of the shaded region.

Solution

Verbal Model Area of shaded region = Area of entire square − Area of white region

Labels Area of shaded region = A **(square units)**

Area of entire square = $(3x + 2)^2$ **(square units)**

Area of white region = $(2x + 3)(2x - 3)$ **(square units)**

Algebraic Model

$A = (3x + 2)^2 - (2x + 3)(2x - 3)$ **Write algebraic model.**

$= (9x^2 + 12x + 4) - (4x^2 - 9)$ **Apply patterns.**

$= 9x^2 + 12x + 4 - 4x^2 + 9$ **Use distributive property.**

$= 5x^2 + 12x + 13$ **Simplify.**

Answer The area of the shaded region is $5x^2 + 12x + 13$ square units.

Checkpoint Complete the following exercise.

3. Write an expression for the area of the shaded region.

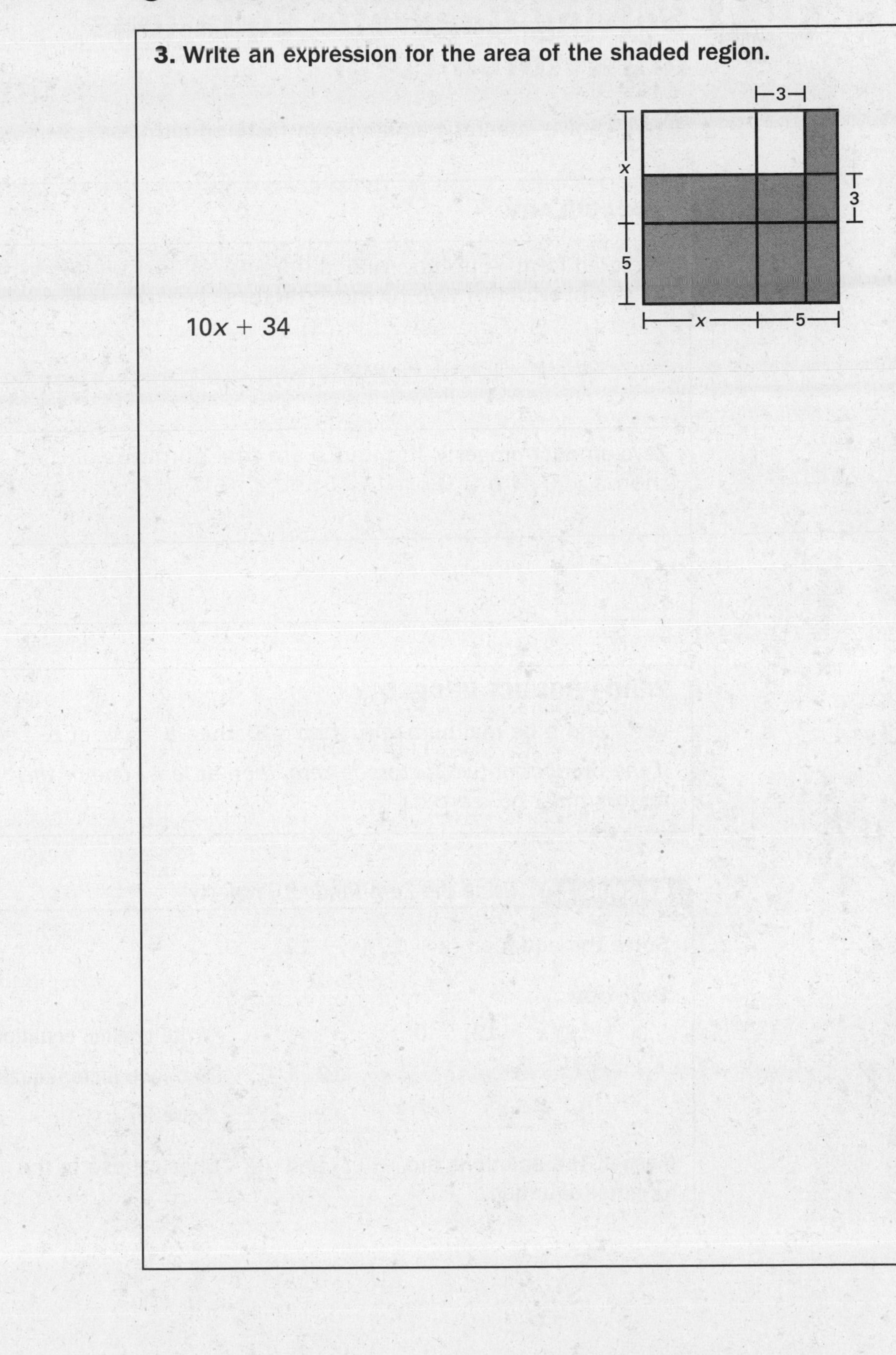

$10x + 34$

10.4 Solving Quadratic Equations in Factored Form

Goal Solve quadratic equations in factored form.

VOCABULARY

Factored form A polynomial is in factored form if it is written as the product of two or more factors.

Zero-product property If a and b are real numbers and $ab = 0$, then $a = 0$ or $b = 0$.

ZERO-PRODUCT PROPERTY

Let a and b be real numbers. If $ab = 0$, then $a =$ 0 or $b =$ 0.

If the product of two factors is zero, then at least one of the factors must be zero.

Example 1 ***Using the Zero-Product Property***

Solve the equation $(x + 17)(x - 12) = 0$.

Solution

$(x + 17)(x - 12) = 0$			**Write original equation.**
$x + 17 = 0$	*or*	$x - 12 = 0$	**Set each factor equal to 0.**
$x = -17$		$x = 12$	**Solve for x.**

Answer The solutions are -17 and 12. Check these in the original equation.

This equation is a square of a binomial, so the factor $(x - 9)$ is a *repeated* factor. Repeated factors are used twice or more in an equation.

Example 2 ***Solve a Repeated-Factor Equation***

Solve $(x - 9)^2 = 0$.

To solve this equation you set $(x - 9)$ equal to zero.

$(x - 9)^2 = 0$	**Write original equation.**
$x - 9 = 0$	**Set factor equal to 0.**
$x = 9$	**Solve for *x*.**

Answer The solution is 9. Check this in the original equation.

Example 3 ***Solve a Factored Cubic Equation***

Solve $(7x + 3)(2x - 1)(x + 5) = 0$.

$(7x + 3)(2x - 1)(x + 5) = 0$ **Original equation**

$7x + 3 = 0$ *or* $2x - 1 = 0$ *or* $x + 5 = 0$ **Set factors equal to 0.**

$7x = -3$ | $2x = 1$ | $x = -5$ **Solve for *x*.**

$x = -\frac{3}{7}$ | $x = \frac{1}{2}$

Answer The solutions are $-\frac{3}{7}$, $\frac{1}{2}$, and -5. Check these in the original equation.

✔ ***Checkpoint*** **Solve the equation and check the solutions.**

1. $(x + 2)(x - 4) = 0$

$-2, 4$

2. $(x - 4)(4x - 8)(3x + 11) = 0$

$4, 2, -\frac{11}{3}$

Example 4 *Graphing a Factored Equation*

Sketch the graph of $y = (x - 1)(x + 3)$.

1. **Find** the x-intercepts. Solve $(x - 1)(x + 3) = 0$ to find the x-intercepts: $\underline{1}$ and $\underline{-3}$.

2. **Use the x-intercepts to find the coordinates of the vertex.**

   - The x-coordinate of the vertex is the average of the x-intercepts.

   $$x = \frac{\boxed{1} + \left(\boxed{-3}\right)}{2} = \frac{\boxed{-2}}{2} = \underline{-1}$$

   - **Substitute** the x-coordinate into the original equation to find the y-coordinate.

   $y = \underline{(-1 - 1)(-1 + 3) = -4}$

   - The vertex is at ($\underline{-1}$, $\underline{-4}$).

3. **Sketch** the graph using the vertex and x-intercepts.

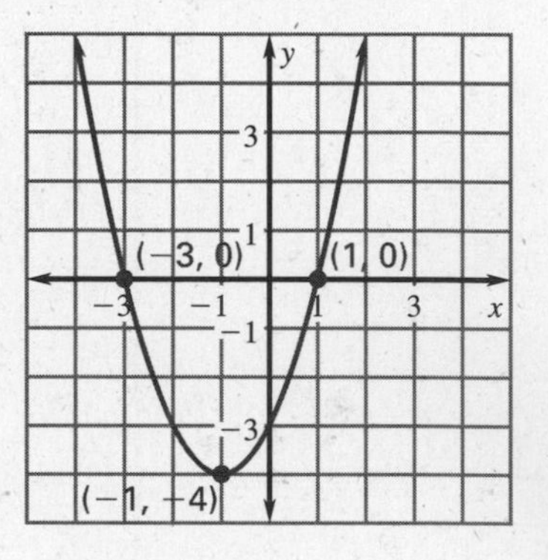

Checkpoint Complete the following exercise.

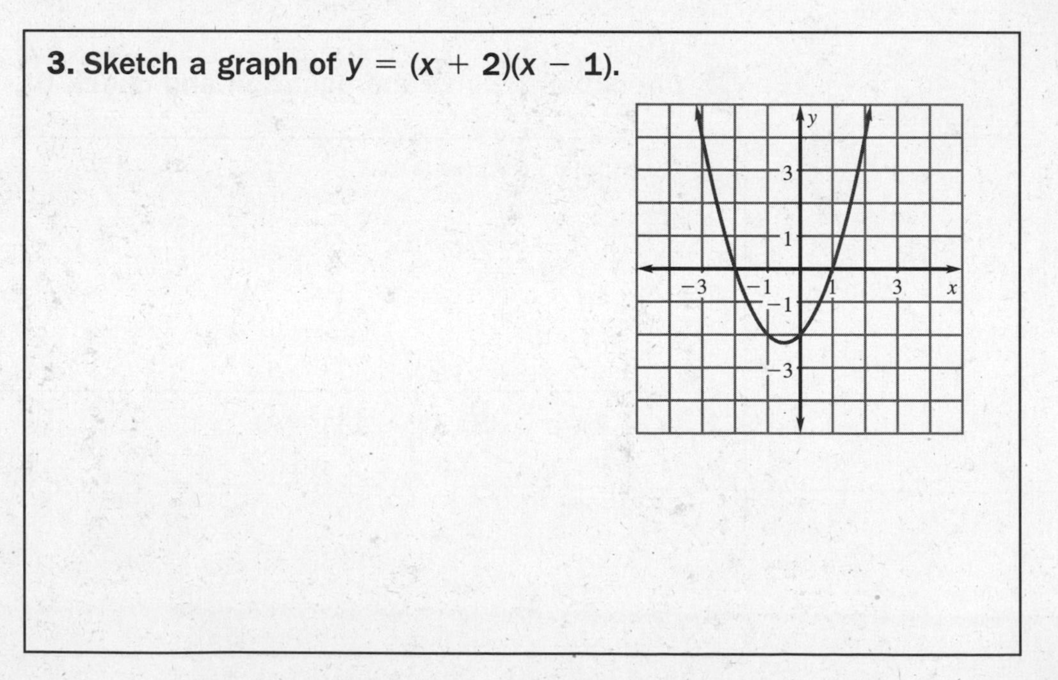

3. Sketch a graph of $y = (x + 2)(x - 1)$.

10.5 Factoring $x^2 + bx + c$

Goal Factor trinomials of the form $x^2 + bx + c$.

FACTORING $x^2 + bx + c$

You know from the FOIL method that

$$(x + p)(x + q) = x^2 + (p + q)x + pq.$$

So to factor $x^2 + bx + c$, you need to find numbers p and q such that $p + q = b$ and $pq = c$.

Example: $x^2 + 6x + 8 = (x + \underline{4})(x + \underline{2})$

$\underline{4} + \underline{2} = 6$ and $\underline{4} \cdot \underline{2} = 8$

Example 1 *Factor when b and c Are Positive*

Factor $x^2 + 5x + 6$.

The first term of each binomial factor is x. For this trinomial, $b = \underline{5}$ and $c = \underline{6}$. You need to find numbers p and q whose product is $\underline{6}$ and whose sum is $\underline{5}$.

p and q	$p + q$
1, 6	7
2, 3	5

The numbers you need are $\underline{2}$ and $\underline{3}$.

Answer $x^2 + 5x + 6 = (\underline{x + 2})(\underline{x + 3})$.

Example 2 *Factor when b is Negative and c is Positive*

Factor $x^2 - 8x + 15$.

The first term of each binomial factor is x. For this trinomial, $b = \underline{-8}$ and $c = \underline{15}$. Because c is positive, you need to find numbers p and q with the $\underline{\text{same}}$ sign. Find numbers p and q whose sum is $\underline{-8}$ and whose product is $\underline{15}$.

p and q	$p + q$
−1, −15	−16
−5, −3	−8

The numbers you need are $\underline{-5}$ and $\underline{-3}$.

Answer $x^2 - 8x + 15 = (\underline{x - 5})(\underline{x - 3})$.

Example 3 ***Factor when b and c Are Negative***

Factor $x^2 - 4x - 5$.

Solution

The first term of each binomial factor is x.

For this trinomial, $b =$ $\underline{-4}$ and $c =$ $\underline{-5}$. Because c is negative, you need to find numbers p and q with $\underline{\text{different}}$ signs. Find numbers p and q whose sum is $\underline{-4}$ and whose product is $\underline{-5}$.

p* and *q	***p* + *q***
-1, $\underline{5}$	$\underline{4}$
1, $\underline{-5}$	$\underline{-4}$

The numbers you need are $\underline{1}$ **and** $\underline{-5}$.

Answer $x^2 - 4x - 5 = (\underline{x + 1})(\underline{x - 5})$.

Example 4 ***Factor when b is Positive and c is Negative***

Factor $x^2 + 8x - 20$.

Solution

The first term of each binomial factor is x.

For this trinomial, $b =$ $\underline{8}$ and $c =$ $\underline{-20}$. Because c is negative, you need to find numbers p and q with $\underline{\text{different}}$ signs. Find numbers p and q whose sum is $\underline{8}$ and whose product is $\underline{-20}$.

As soon as you find the correct pair of numbers for a trinomial, you can stop listing all possible pairs. For example, in Example 4 you do not need to list the pairs 4 and -5, or -4 and 5.

p* and *q	***p* + *q***
1, $\underline{-20}$	$\underline{-19}$
-1, $\underline{20}$	$\underline{19}$
2, $\underline{-10}$	$\underline{-8}$
-2, $\underline{10}$	$\underline{8}$

The numbers you need are $\underline{-2}$ **and** $\underline{10}$.

Answer $x^2 + 8x - 20 = (\underline{x - 2})(\underline{x + 10})$.

Example 5 *Solve a Quadratic Equation*

Solve $x^2 + 9x = -14$ by factoring.

$x^2 + 9x = -14$	**Write equation.**
$x^2 + 9x \underline{+ 14} = 0$	**Write in standard form.**
$\underline{(x + 2)(x + 7)} = 0$	**Factor left side.**
$\underline{x + 2} = 0$ *or* $\underline{x + 7} = 0$	**Use zero-product property.**
$x = \underline{-2}$ | $x = \underline{-7}$	**Solve for *x*.**

Answer The solutions are $\underline{-2}$ and $\underline{-7}$.

Checkpoint **Factor the trinomial.**

1. $x^2 + 11x + 18$

$(x + 2)(x + 9)$

2. $x^2 - 10x + 21$

$(x - 3)(x - 7)$

3. $x^2 - 6x - 7$

$(x + 1)(x - 7)$

4. $x^2 + 3x - 10$

$(x - 2)(x + 5)$

5. Solve the equation $x^2 + 4x - 21 = 0$ by factoring.

$-7, 3$

10.6 Factoring $ax^2 + bx + c$

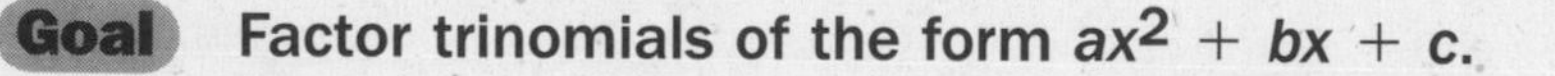

Goal Factor trinomials of the form $ax^2 + bx + c$.

Example 1 *Factor When a and c Are Prime Numbers*

Factor $3x^2 + 22x + 7$.

1. Write the numbers m and n whose product is 3 and the numbers p and q whose product is 7.

m and *n*	*p* and *q*
1, 3	1, 7

2. Use these numbers to write trial factors. Then use the **O**uter and **I**nner products of **FOIL** to check the middle term.

Trial Factors	Middle Term
$(1x + 1)(3x + 7)$	$7x + 3x = 10x$
$(3x + 1)(1x + 7)$	$21x + 1x = 22x$

Answer $3x^2 + 22x + 7 = (3x + 1)(x + 7)$.

Example 2 *Factor When a and c Are Not Prime Numbers*

Factor $4x^2 - 20x + 9$.

For this trinomial, $a = 4$, $b = -20$, and $c = 9$. Because c is positive, you need to find numbers p and q with the same sign. Because b is negative, only negative numbers p and q need to be tried.

1. Write the numbers m and n whose product is 4 and the numbers p and q whose product is 9.

m and *n*	*p* and *q*
1, 4	−1, −9
2, 2	−3, −3

2. Use these numbers to write trial factors. Then use the **O**uter and **I**nner products of **FOIL** to check the middle term.

Trial Factors	Middle Term
$(1x - 1)(4x - 9)$	$-9x - 4x = -13x$
$(1x - 9)(4x - 1)$	$-1x - 36x = -37x$
$(2x - 1)(2x - 9)$	$-18x - 2x = -20x$

Answer $4x^2 - 20x + 9 = (2x - 1)(2x - 9)$.

Checkpoint Factor the trinomial.

1. $7x^2 + 12x + 5$	2. $15x^2 - 13x + 2$
$(x + 1)(7x + 5)$	$(3x - 2)(5x - 1)$

Example 3 ***Factor with a Common Factor for a, b, and c***

Factor $9x^2 + 42x - 15$.

The coefficients of this trinomial have a common factor of 3.

$3(3x^2 + 14x - 5)$ **Factor out the common factor.**

It remains to factor a trinomial with $a = 3$, $b = 14$, and $c = -5$. Because c is negative, you need to find numbers p and q with different signs.

1. **Write the numbers m and n whose product is 3 and the numbers p and q whose product is -5.**

m and *n*	*p* and *q*
1, 3	−1, 5
	1, −5

2. **Use these numbers to write trial factors. Then use the Outer and Inner products of FOIL to check the middle term.**

Trial Factors	Middle Term
$(1x - 1)(3x + 5)$	$5x - 3x = 2x$
$(1x + 5)(3x - 1)$	$-x + 15x = 14x$

Remember to include the common factor 3 in the complete factorization.

Answer $9x^2 + 42x - 15 = 3(x + 5)(3x - 1)$.

Example 4 ***Solve a Quadratic Equation***

$14n^2 + 10n + 2 = -17n - 7$	**Original equation**
$\underline{14n^2} + \underline{27n} + \underline{9} = 0$	**Write in standard form.**
$(\underline{7n + 3})(\underline{2n + 3}) = 0$	**Factor left side.**
$\underline{7n + 3} = 0$ *or* $\underline{2n + 3} = 0$	**Use zero-product property.**
$n = \underline{-\frac{3}{7}}$ | $n = \underline{-\frac{3}{2}}$	**Solve for *n*.**

Answer The solutions are $\underline{-\frac{3}{7}}$ and $\underline{-\frac{3}{2}}$. Check these in the original equation.

✔ ***Checkpoint*** **Factor the trinomial.**

3. $15x^2 - 35x + 10$ $5(3x - 1)(x - 2)$	**4.** $-12x^2 + 10x + 2$ $-2(x - 1)(6x + 1)$

Solve the quadratic equation.

5. $12n^2 - 21n + 2 = -8n + 6$

$-\frac{1}{4}, \frac{4}{3}$

10.7 Factoring Special Products

Goal Factor special products.

VOCABULARY

Perfect square trinomial A perfect square trinomial is a trinomial of the form $a^2 + 2ab + b^2$ and $a^2 - 2ab + b^2$; perfect square trinomials can be factored as the squares of binomials.

FACTORING SPECIAL PRODUCTS

Difference of Two Squares Patterns

$a^2 - b^2 = \underline{(a + b)(a - b)}$

Example:

$9x^2 - 25 = (\underline{3x} + \underline{5})(\underline{3x} - \underline{5})$

Perfect Square Trinomial Pattern

$a^2 + 2ab + b^2 = \underline{(a + b)^2}$

$a^2 - 2ab + b^2 = \underline{(a - b)^2}$

Example:

$x^2 + 14x + 49 = (\underline{x} + \underline{7})^2$

$x^2 - 12x + 36 = (\underline{x} - \underline{6})^2$

Example 1 ***Factor the Difference of Two Squares***

a. $m^2 - 49$ **b.** $9x^2 - 16$ **c.** $16p^2 - 25$ **d.** $x^2 - 27$

Solution

a. $m^2 - 49 = m^2 - \underline{7}^2$ **Write as $a^2 - b^2$.**

$= (\underline{m + 7})(\underline{m - 7})$ **Factor using pattern.**

b. $9x^2 - 16 = (\underline{3x})^2 - \underline{4}^2$ **Write as $a^2 - b^2$.**

$= (\underline{3x + 4})(\underline{3x - 4})$ **Factor using pattern.**

c. $16p^2 - 25 = (\underline{4p})^2 - \underline{5}^2$ **Write as $a^2 - b^2$.**

$= (\underline{4p + 5})(\underline{4p - 5})$ **Factor using pattern.**

d. $x^2 - 27$ **cannot be factored using integers because** it does not fit the difference of two squares pattern**; 27 is not the** square of an integer**.**

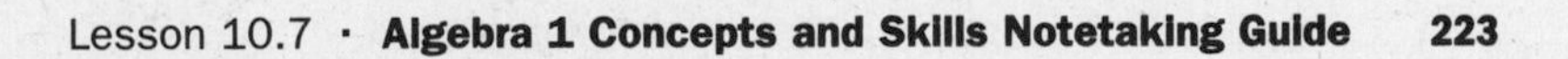

Example 2 *Factor Perfect Square Trinomials*

a. $x^2 + 10x + 25$

$= x^2 + 2(\underline{x})(\underline{5}) + \underline{5}^2$ **Write as $a^2 + 2ab + b^2$.**

$= (\underline{x + 5})^2$ **Factor using pattern.**

b. $c^2 + 8c + 16$

$= c^2 + 2(\underline{c})(\underline{4}) + \underline{4}^2$ **Write as $a^2 + 2ab + b^2$.**

$= (\underline{c + 4})^2$ **Factor using pattern.**

c. $9y^2 - 12y + 4$

$= (\underline{3y})^2 - 2(\underline{3y})(\underline{2}) + \underline{2}^2$ **Write as $a^2 - 2ab + b^2$.**

$= (\underline{3y - 2})^2$ **Factor using pattern.**

✔ ***Checkpoint*** **Factor the expression.**

1. $t^2 - 4$ $(t + 2)(t - 2)$	**2.** $49y^2 - 25$ $(7y + 5)(7y - 5)$
3. $4a^2 + 12a + 9$ $(2a + 3)^2$	**4.** $y^2 - 18y + 81$ $(y - 9)^2$

Example 3 *Factor Out a Constant First*

a. $27 - 75x^2 = 3(\underline{9} - \underline{25x^2})$ **Factor out common factor.**

$= 3\left[\underline{3}^2 - (\underline{5x})^2\right]$ **Write as $a^2 - b^2$.**

$= 3(\underline{3 + 5x})(\underline{3 - 5x})$ **Factor using pattern.**

b. $2c^2 + 8c + 8$

$= 2(\underline{c^2} + \underline{4c} + \underline{4})$ **Factor out common factor.**

$= 2(\underline{c^2} + 2(\underline{c})(\underline{2}) + \underline{2}^2)$ **Write as $a^2 + 2ab + b^2$.**

$= 2(\underline{c + 2})^2$ **Factor using pattern.**

c. $5y^2 - 35y + 90 = 5(\underline{y^2 - 7y + 18})$ **Factor out common factor.**

Example 4 ***Solve a Quadratic Equation***

Solve $9x^2 + 6x + 1 = 0$.

Solution

$9x^2 + 6x + 1 = 0$	**Write original equation.**
$(\underline{3x})^2 + 2(\underline{3x})(\underline{1}) + \underline{1}^2 = 0$	**Write as $a^2 + 2ab + b^2$.**
$(\underline{3x + 1})^2 = 0$	**Factor using pattern.**
$(\underline{3x + 1}) = \underline{0}$	**Set repeated factor equal to $\underline{0}$.**
$x = \underline{-\frac{1}{3}}$	**Solve for x.**

Answer The solution is $\underline{-\frac{1}{3}}$. Check this in the original equation.

Checkpoint **Factor the expression.**

5. $20z^2 - 45$ $5(2z + 3)(2z - 3)$	**6.** $4y^2 - 40y + 100$ $4(y - 5)^2$

Solve the equation by factoring.

7. $81x^2 - 16 = 0$ $-\frac{4}{9}, \frac{4}{9}$	**8.** $3w^2 - 36w + 108 = 0$ 6

Factoring Cubic Polynomials

Goal Factor cubic polynomials.

VOCABULARY

Prime polynomial A polynomial is prime if it cannot be factored using integer coefficients.

Factor a polynomial completely To factor a polynomial completely, write it as the product of monomial and prime factors.

Example 1 ***Factor Completely***

Factor $6x^4 - 18x^3 + 12x^2$ completely.

Solution

First find the greatest common factor of $6x^4$, $18x^3$, and $12x^2$.

$6x^4 = 2 \cdot 3 \cdot x \cdot x \cdot x \cdot x$

$18x^3 = 2 \cdot 3 \cdot 3 \cdot x \cdot x \cdot x$

$12x^2 = 2 \cdot 2 \cdot 3 \cdot x \cdot x$

GCF = $2 \cdot 3 \cdot x \cdot x = 6x^2$

$6x^4 - 18x^3 + 12x^2 = 6x^2(x^2 - 3x + 2)$ **Factor out GCF.**

$= 6x^2(x - 2)(x - 1)$ **Factor trinomial.**

Monomial factor: $6x^2$ **Prime factors:** $(x - 2)(x - 1)$

Answer $6x^4 - 18x^3 + 12x^2 = 6x^2(x - 2)(x - 1)$.

Example 2 *Factor by Grouping*

Factor $5x^3 + 2x^2 + 5x + 2$ completely.

Solution

$$5x^3 + 2x^2 + 5x + 2$$

$$= (5x^3 + \underline{5x}) + (\underline{2x^2} + 2)$$ **Group terms.**

$$= \underline{5x}(\underline{x^2} + \underline{1}) + \underline{2}(\underline{x^2} + \underline{1})$$ **Factor each group.**

$$= (\underline{x^2} + \underline{1})(\underline{5x} + \underline{2})$$ **Use distributive property.**

✓ ***Checkpoint*** **Factor the expression completely.**

1. $3x^3 - 18x^2 + 12x$

$3x(x^2 - 6x + 4)$

2. $4x^3 - 16x$

$4x(x + 2)(x - 2)$

3. $2x^3 + 8x^2 + 5x + 20$

$(x + 4)(2x^2 + 5)$

4. $x^3 + 2x^2 - 9x - 18$

$(x + 2)(x + 3)(x - 3)$

FACTORING MORE SPECIAL PRODUCTS

Sum of Two Cubes Pattern

$a^3 + b^3 = (\underline{\ a + b\ })(\underline{\ a^2 - ab + b^2\ })$

Example: $(x^3 + 1) = (\underline{\ x + 1\ })(\underline{\ x^2 - x + 1\ })$

Difference of Two Cubes Pattern

$a^3 - b^3 = (\underline{\ a - b\ })(\underline{\ a^2 + ab + b^2\ })$

Example: $(x^3 - 8) = (\underline{\ x - 2\ })(\underline{\ x^2 + 2x + 4\ })$

Example 3 *Factor the Sum or Difference of Two Cubes*

Factor the expression.

a. $x^3 + 64$

b. $x^3 - 27$

Solution

a. $x^3 + 64 = x^3 + \underline{\ 4\ }^3$ **Write as sum of cubes.**

$= (x + \underline{\ 4\ })(\underline{\ x^2\ } - \underline{\ 4x\ } + \underline{\ 16\ })$ **Use special product pattern.**

b. $x^3 - 27 = x^3 - \underline{\ 3\ }^3$ **Write as difference of cubes.**

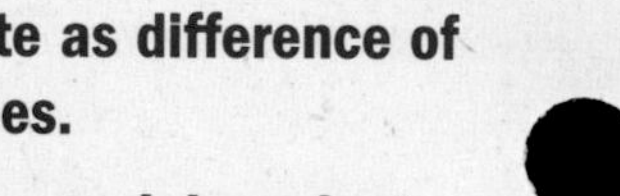

$= (x - \underline{\ 3\ })(\underline{\ x^2\ } + \underline{\ 3x\ } + \underline{\ 9\ })$ **Use special product pattern.**

✔ ***Checkpoint*** **Factor the expression.**

5. $x^3 - 1$	**6.** $x^3 + 125$
$(x - 1)(x^2 + x + 1)$	$(x + 5)(x^2 - 5x + 25)$

Words to Review

Give an example of the vocabulary word.

Monomial $7x^3$	**Degree of a monomial** The degree of $5x^2y$ is $2 + 1 = 3$.
Polynomial $-3x^5 + x^3 + 2 - x^2$	**Binomial** $3x^5 + x^3$
Trinomial $x^3 - x^2 + 2$	**Standard form of a polynomial** $-3x^5 + x^3 - x^2 + 2$
Degree of a polynomial The degree of $3x^5 + x^3 - x^2 + 2$ is 5.	**FOIL pattern** $(x - 3)(x + 1) = x^2 + x - 3x - 3$

Factored form	**Zero-product property**
$(x + 4)(x - 4) = 0$	If $(x + 4)(x - 4) = 0$, then $x + 4 = 0$ or $x - 4 = 0$.
Factor a trinomial	**Perfect square trinomial**
$x^2 + 2x - 8 = (x + 4)(x - 2)$	$x^2 + 6x + 9$
Prime polynomial	**Factor a polynomial completely**
$x^2 + 4$	$3x^2 - 27 = 3(x^2 - 9)$ $= 3(x - 3)(x + 3)$

Review your notes and Chapter 10 by using the Chapter Review on pages 623–626 of your textbook.

11.1 Proportions

Goal Solve proportions.

VOCABULARY

Proportion A proportion is an equation that states that two ratios are equal.

Extremes In the proportion $\frac{a}{b} = \frac{c}{d}$, a and d are the extremes.

Means In the proportion $\frac{a}{b} = \frac{c}{d}$, b and c are the means.

RECIPROCAL PROPERTY OF PROPORTIONS

If two ratios are equal, their reciprocals are also equal.

If $\frac{a}{b} = \frac{c}{d}$, then $\frac{b}{a} = \frac{d}{c}$.

Example: $\frac{2}{3} = \frac{4}{6} \rightarrow \frac{3}{2} = \frac{6}{4}$

Example 1 ***Use the Reciprocal Property***

Solve the proportion $\frac{4}{y} = \frac{2}{7}$ using the reciprocal property.

1. Write the original proportion. $\frac{4}{y} = \frac{2}{7}$
2. Use the reciprocal property. $\frac{y}{4} = \frac{7}{2}$
3. Multiply each side of the equation by 4 to clear the equation of fractions. $y = 14$

Answer The solution is $y =$ 14. Check this in the original equation.

CROSS PRODUCT PROPERTY OF PROPORTIONS

The product of the extremes equals the product of the means.

If $\frac{a}{b} = \frac{c}{d}$, then $ad = bc$. **Example:** $\frac{2}{3} = \frac{4}{6} \Rightarrow 2 \cdot 6 = 3 \cdot 4$

Example 2 *Use the Cross Product Property*

Solve $\frac{x}{16} = \frac{5}{4}$ using the cross product property.

1. **Write** the original proportion. $\frac{x}{16} = \frac{5}{4}$
2. **Use** the cross product property. $x \cdot 4 = 16 \cdot 5$
3. **Simplify** the equation. $4x = 80$
4. **Solve** by dividing each side by 4. $x = 20$

Check Substituting 20 for x, $\frac{20}{16}$ simplifies to $\frac{5}{4}$.

Example 3 *Use the Cross Product Property*

Solve the proportion $\frac{x+1}{10} = \frac{2}{x}$.

1. **Write** the original proportion. $\frac{x+1}{10} = \frac{2}{x}$
2. **Use** the cross product property. $(x+1)x = 10(2)$
3. **Multiply.** $x^2 + x = 20$
4. **Collect** terms on one side. $x^2 + x - 20 = 0$
5. **Factor** the left-hand side. $(x-4)(x+5) = 0$
6. **Solve** the equation. $x = 4$ or -5

Answer The solutions are $x = 4$ and $x = -5$. Check both solutions.

Remember to check your solution in the *original* proportion. Since Example 3 has two solutions, you need to check both of them.

Example 4 *Cross Multiply and Check Solutions*

Solve the equation $\dfrac{x^2 - 16}{x + 4} = \dfrac{x - 4}{3}$.

1. **Write** the original proportion. $\dfrac{x^2 - 16}{x + 4} = \dfrac{x - 4}{3}$
2. **Cross multiply.** $(\underline{x^2 - 16})(\underline{3}) = (\underline{x + 4})(\underline{x - 4})$
3. **Multiply.** $\underline{3x^2 - 48} = \underline{x^2 - 16}$
4. **Isolate** the variable term. $x^2 = \underline{16}$
5. **Solve** by taking the square root of each side. $x = \underline{\pm 4}$

The solutions appear to be $x = \underline{4}$ and $x = \underline{-4}$. However, you must discard $x = \underline{-4}$, since the denominator on the left-hand side would become zero.

Answer The solution is $x = \underline{4}$.

Checkpoint **Solve the proportion. Check your solutions.**

1. $\dfrac{35}{y} = \dfrac{5}{7}$ 49	**2.** $\dfrac{3}{8} = \dfrac{2}{x}$ $\dfrac{16}{3}$	**3.** $\dfrac{4}{7} = \dfrac{2u}{5}$ $\dfrac{10}{7}$
4. $\dfrac{v + 2}{4} = \dfrac{v}{3}$ 6	**5.** $\dfrac{3}{y} = \dfrac{2y + 1}{5}$ $-3, \dfrac{5}{2}$	**6.** $\dfrac{m}{4m} = \dfrac{2m - 1}{3}$ $\dfrac{7}{8}$

11.2 Direct and Inverse Variation

Goal Use direct and inverse variation.

VOCABULARY

Inverse variation Inverse variation is the relationship between two variables x and y for which there is a nonzero number k such that $xy = k$ or $y = \frac{k}{x}$. The variables x and y are said to vary inversely.

MODELS FOR DIRECT AND INVERSE VARIATION

Direct Variation

The variables x and y vary directly if for a constant k

$$\frac{y}{x} = k, \text{ or } y = kx, \text{ where } k \neq 0.$$

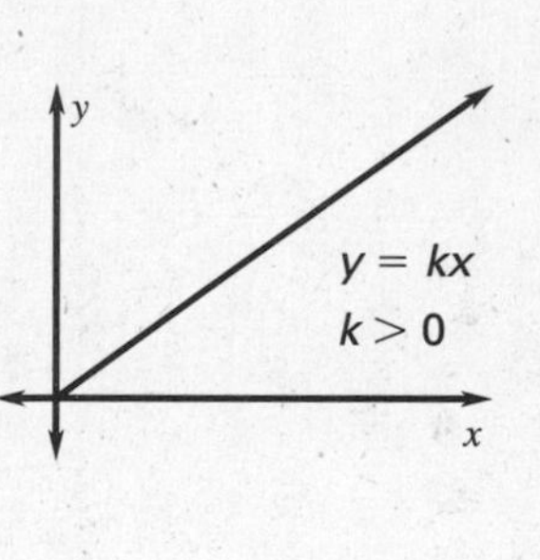

Inverse Variation

The variables x and y vary inversely if for a constant k

$$xy = k, \text{ or } y = \frac{k}{x}, \text{ where } k \neq 0.$$

The number k is the constant of variation.

$y = \frac{k}{x}$

$k > 0$

Example 1 ***Use Direct Variation***

Find an equation that relates x and y such that x and y vary directly, and $y = 8$ when $x = 2$.

1. Write the direct variation model. $\frac{y}{x} = k$

2. Substitute 8 for y and 2 for x. $\frac{8}{2} = k$

3. Simplify the left-hand side. $4 = k$

Answer The direct variation is $\frac{y}{x} = 4$, or $y = 4x$.

Example 2 ***Use Inverse Variation***

Find an equation that relates x and y such that x and y vary inversely, and $y = 8$ when $x = 2$.

1. Write the inverse variation model. $xy = k$

2. Substitute 8 for y and 2 for x. $(2)(8) = k$

3. Simplify the left-hand side. $16 = k$

Answer The inverse variation that relates x and y is $xy = 16$, or $y = \frac{16}{x}$.

Checkpoint **Complete the following exercises.**

1. Find an equation that relates x and y such that x and y vary directly, and $x = 24$ and $y = 6$.

$y = \frac{1}{4}x$

2. Find an equation that relates x and y such that x and y vary inversely, and $x = 4$ and $y = 2$.

$y = \frac{8}{x}$

Example 3 *Compare Direct and Inverse Variation*

Direct and inverse variation models represent functions because for each value of x there is exactly one value of y. For inverse variation, the domain excludes 0.

Compare the direct variation model and the inverse variation model you found in Examples 1 and 2 using $x = -4, -3, -2, -1, 1, 2, 3,$ and 4.

a. numerically **b.** graphically

Solution

a. Use the models $y = 4x$ and $y = \frac{16}{x}$ to make a table.

x-value	-4	-3	-2	-1	1	2	3	4
Direct, $y = 4x$	-16	-12	-8	-4	4	8	12	16
Inverse, $y = \frac{16}{x}$	-4	$-\frac{16}{3}$	-8	-16	16	8	$\frac{16}{3}$	4

Direct Variation: Because k is positive, y increases as x increases. As x increases by 1, y increases by 4.

Inverse Variation: Because k is positive, y decreases as x increases.

b. Use the table of values to graph each model.

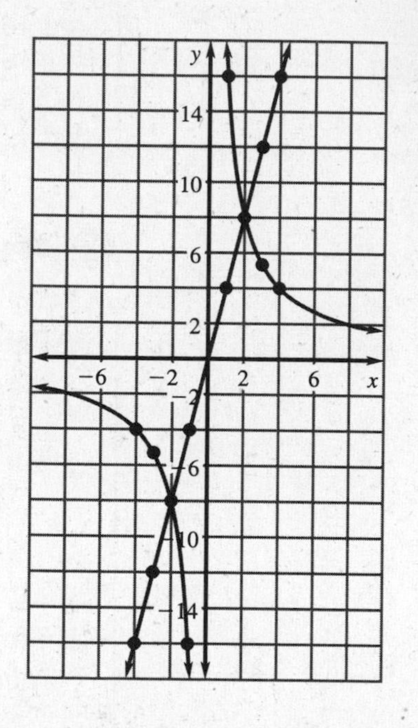

Direct Variation: The graph of this model is a line passing through the origin.

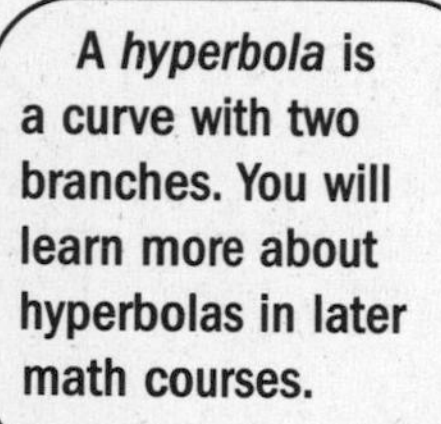

Inverse Variation: The graph for this model is a *hyperbola*. Since neither x nor y can equal 0, the graph does not intersect either axis.

11.3 Simplifying Rational Expressions

Goal Simplify rational expressions.

VOCABULARY

Rational number A rational number is a number that can be written as the quotient of two integers.

Rational expression A rational expression is a fraction whose numerator and denominator are nonzero polynomials.

SIMPLIFYING RATIONAL EXPRESSIONS

Let a, b, and c be nonzero numbers.

$$\frac{ac}{bc} = \frac{a \cdot \cancel{c}}{b \cdot \cancel{c}} = \frac{a}{b}$$

Example 1 ***Simplify Rational Expressions***

Simplify the rational expression if possible.

a. $\frac{8x}{2} = \frac{\cancel{2} \cdot 2 \cdot 2 \cdot x}{\cancel{2}} = 4x$

b. $\frac{15x^2}{10x} = \frac{3 \cdot \cancel{5} \cdot \cancel{x} \cdot x}{2 \cdot \cancel{5} \cdot \cancel{x}} = \frac{3x}{2}$

Example 2 *Write in Simplest Form*

Simplify $\frac{x^2}{x(x^2 + 1)}$.

$$\frac{x^2}{x(x^2+1)} = \frac{\boxed{\cancel{x}} \cdot \boxed{x}}{\boxed{\cancel{x}} \cdot (x^2+1)}$$ Divide out the common factor x.

$$= \frac{\boxed{x}}{\boxed{x^2+1}}$$ Simplify.

Example 3 *Factor Numerator and Denominator*

Simplify $\frac{6x^3}{3x - 12x^2}$.

1. Write the original expression. $\frac{6x^3}{3x - 12x^2}$
2. Factor the numerator and denominator. $\frac{2 \cdot 3 \cdot x \cdot x \cdot x}{3x(1 - 4x)}$
3. Divide out the common factors 3 and x. $\frac{\cancel{3x}(2x^2)}{\cancel{3x}(1 - 4x)}$
4. Simplify the expression. $\frac{2x^2}{1 - 4x}$

✔ *Checkpoint* Simplify the expression if possible.

1. $\frac{4x}{2x^2}$	2. $\frac{x + 1}{2x}$	3. $\frac{3x^2}{9x - 15x^2}$
$\frac{2}{x}$	Already in simplest form	$\frac{x}{3 - 5x}$

Example 4 ***Recognize Opposite Factors***

1. Write the original expression. $\dfrac{9 - x^2}{x^2 - 4x + 3}$

2. Factor the numerator and denominator. $\dfrac{(\boxed{3 - x})(3 + x)}{(\boxed{x - 3})(x - 1)}$

3. Factor -1 **from** ($3 - x$). $\dfrac{-(\boxed{x - 3})(3 + x)}{(\boxed{x - 3})(x - 1)}$

4. Divide out the common factor ($x - 3$). $\dfrac{-(\boxed{\cancel{x - 3}})(3 + x)}{(\boxed{\cancel{x - 3}})(x - 1)}$

5. Simplify the expression. $-\dfrac{3 + x}{x - 1}$

Example 5 ***Divide a Polynomial by a Binomial***

Divide $(x^2 - 3x - 10)$ **by** $(x - 5)$.

1. Rewrite the problem as a rational expression. $\dfrac{x^2 - 3x - 10}{x - 5}$

2. Factor the numerator. $\dfrac{(x - 5)(x + 2)}{x - 5}$

3. Divide out the common factor ($x - 5$). $\dfrac{\cancel{(x - 5)}(x + 2)}{\cancel{x - 5}}$

4. Simplify the expression. $x + 2$

Checkpoint **Simplify the expression.**

4. $\dfrac{3(5 - x)}{12(x - 5)}$	**5.** $\dfrac{x^2 + 8x + 15}{x + 3}$
$-\dfrac{1}{4}$	$x + 5$

11.4 Multiplying and Dividing Rational Expressions

Goal Multiply and divide rational expressions.

MULTIPLYING AND DIVIDING RATIONAL EXPRESSIONS

Let a, b, c, and d be nonzero polynomials.

To multiply, multiply the numerators and denominators.

$$\frac{a}{b} \cdot \frac{c}{d} = \frac{ac}{bd}$$

To divide, multiply by the reciprocal of the divisor.

$$\frac{a}{b} \div \frac{c}{d} = \frac{a}{b} \cdot \frac{d}{c}$$

Example 1 ***Multiply Rational Expressions***

Simplify $\frac{3x}{2x^2 - 4x} \cdot \frac{x - 2}{3x^2 + 5x - 2}$.

1. **Write** the original expression. $\frac{3x}{2x^2 - 4x} \cdot \frac{x - 2}{3x^2 + 5x - 2}$

2. **Factor** the numerators and denominators. $\frac{3x}{2x(x - 2)} \cdot \frac{x - 2}{(3x - 1)(x + 2)}$

3. **Multiply** the numerators and denominators. $\frac{3x(x - 2)}{2x(x - 2)(3x - 1)(x + 2)}$

4. **Divide out** the common factors. $\frac{3\cancel{x}\cancel{(x - 2)}}{2\cancel{x}\cancel{(x - 2)}(3x - 1)(x + 2)}$

5. **Simplify** the expression. $\frac{3}{2(3x - 1)(x + 2)}$

Example 2 *Multiply by a Polynomial*

Simplify $\dfrac{8x}{x^2 - 7x + 12} \cdot (x \quad 3)$.

Solution

$\dfrac{8x}{x^2 - 7x + 12} \cdot (x - 3)$	**Write original expression.**
$= \dfrac{8x}{x^2 - 7x + 12} \cdot \dfrac{x - 3}{1}$	**Write $x - 3$ as $\dfrac{x - 3}{1}$.**
$= \dfrac{8x}{(x - 4)(x - 3)} \cdot \dfrac{x - 3}{1}$	**Factor.**
$= \dfrac{8x(x - 3)}{(x - 4)(x - 3)}$	**Multiply numerators and denominators.**
$= \dfrac{8x\cancel{(x - 3)}}{(x - 4)\cancel{(x - 3)}}$	**Divide out common factor.**
$= \dfrac{8x}{(x - 4)}$	**Simplify the expression.**

Checkpoint **Simplify the expression.**

1. $\dfrac{3y^2}{4y^3} \cdot \dfrac{16y^4}{9y^5}$ $\dfrac{4}{3y^2}$	**2.** $\dfrac{5m}{3m^2 + 9m} \cdot (m + 3)$ $\dfrac{5}{3}$

Example 3 ***Divide Rational Expressions***

Simplify $\frac{n + 7}{n} \div \frac{n + 7}{n - 1}$.

1. Write the original expression. $\frac{n + 7}{n} \div \frac{n + 7}{n - 1}$

2. Multiply by the reciprocal. $\frac{n + 7}{n} \cdot \frac{n - 1}{n + 7}$

3. Multiply the numerators and denominators. $\frac{(n + 7)(n - 1)}{n(n + 7)}$

4. Divide out common factor ($n + 7$). $\frac{\cancel{(n + 7)}(n - 1)}{n\cancel{(n + 7)}}$

5. Simplify the expression. $\frac{n - 1}{n}$

Example 4 ***Divide by a Polynomial***

Simplify $\frac{x^2 - 36}{4x^2} \div (x - 6)$.

Solution

$\frac{x^2 - 36}{4x^2} \div (x - 6) = \frac{x^2 - 36}{4x^2} \cdot \frac{1}{x - 6}$ **Multiply by reciprocal.**

$= \frac{(x + 6)(x - 6)}{4x^2} \cdot \frac{1}{x - 6}$ **Factor.**

$= \frac{(x + 6)(x - 6)}{4x^2(x - 6)}$ **Multiply numerators and denominators.**

$= \frac{(x + 6)\cancel{(x - 6)}}{4x^2\cancel{(x - 6)}}$ **Divide out common factor.**

$= \frac{x + 6}{4x^2}$ **Write in simplest form.**

11.5 Adding and Subtracting with Like Denominators

Goal Add and subtract rational expressions with like denominators.

ADDING OR SUBTRACTING WITH LIKE DENOMINATORS

Let a, b, and c be polynomials, with $c \neq 0$.

To add, add the numerators. $\frac{a}{c} + \frac{b}{c} = \underline{\frac{a + b}{c}}$

To subtract, subtract the numerators. $\frac{a}{c} - \frac{b}{c} = \underline{\frac{a - b}{c}}$

Example 1 ***Add Rational Expressions***

1. Write the original expression. $\frac{7}{x} + \frac{2x - 7}{x}$
2. Add the numerators. $\frac{\boxed{7} + (\boxed{2x - 7})}{x}$
3. Combine like terms. $\underline{\frac{2x}{x}}$
4. Simplify the expression. $\underline{2}$

Example 2 ***Subtract Rational Expressions***

1. Write the original expression. $\frac{5}{3m - 4} - \frac{2m + 1}{3m - 4}$
2. Subtract the numerators. $\frac{\boxed{5} - (\boxed{2m + 1})}{3m - 4}$
3. Distribute the negative. $\frac{\boxed{5 - 2m - 1}}{3m - 4}$
4. Simplify the numerator. $\underline{\frac{4 - 2m}{3m - 4}}$

Example 3 *Simplify After Subtracting*

Simplify $\frac{3x}{2x^2 + 3x - 2} - \frac{x + 1}{2x^2 + 3x - 2}$.

$$\frac{3x}{2x^2 + 3x - 2} - \frac{x + 1}{2x^2 + 3x - 2}$$ Write original expression.

$$= \frac{\boxed{3x} - \left(\boxed{x + 1}\right)}{2x^2 + 3x - 2}$$ Subtract numerators.

$$= \frac{\boxed{2x - 1}}{2x^2 + 3x - 2}$$ Simplify.

$$= \frac{\boxed{2x - 1}}{\left(\boxed{2x - 1}\right)\left(\boxed{x + 2}\right)}$$ Factor.

$$= \frac{\cancel{2x - 1}}{(\cancel{2x - 1})(x + 2)}$$ Divide out common factor.

$$= \frac{1}{x + 2}$$ Write simplest form.

Checkpoint Simplify the expression.

1. $\frac{5}{3x} + \frac{x - 6}{3x}$

   $\frac{x - 1}{3x}$

2. $\frac{9}{2n - 1} - \frac{4n}{2n - 1}$

   $\frac{9 - 4n}{2n - 1}$

3. $\frac{5x}{x^2 - 11x + 28} - \frac{3x + 14}{x^2 - 11x + 28}$

   $\frac{2}{(x - 4)}$

11.6 Adding and Subtracting with Unlike Denominators

Add and subtract rational expressions with unlike denominators.

VOCABULARY

Least common denominator (LCD) The least common denominator or LCD is the least common multiple of the denominators of two or more fractions.

Example 1 ***Find the LCD of Rational Expressions***

Find the least common denominator of $\frac{x-1}{18x}$ and $\frac{-7}{24x^2}$.

1. **Factor** the denominators. $18x = \underline{2 \cdot 3^2 \cdot x}$
   $24x^2 = \underline{2^3 \cdot 3 \cdot x^2}$
2. **Find the highest power of each factor that appears in either denominator.** $\underline{2^3}$, $\underline{3^2}$, $\underline{x^2}$
3. **Multiply** these to find the LCD. $\underline{2^3 \cdot 3^2 \cdot x^2} = \underline{72x^2}$

Answer The LCD is $\underline{72x^2}$.

Example 2 ***Rewrite Rational Expressions***

Find the missing numerator.

$$\frac{2x-1}{2x^3} = \frac{?}{10x^5}$$

Multiply $2x^3$ by $\underline{5x^2}$ to get $10x^5$.

$$\frac{2x-1}{2x^3} = \frac{\boxed{(2x-1) \cdot 5x^2}}{10x^5}$$

Therefore, multiply $\underline{(2x-1)}$ by $\underline{5x^2}$ to get $\underline{(2x-1) \cdot 5x^2}$.

$$\frac{2x-1}{2x^3} = \frac{\boxed{10x^3 - 5x^2}}{10x^5}$$

Simplify.

Example 3 *Add with Unlike Denominators*

Simplify $\frac{3}{x^3} + \frac{3 - x}{x^4}$.

1. Find the LCD. The LCD is x^4.

2. Write the original expression. $\frac{3}{x^3} + \frac{3 - x}{x^4}$

3. Rewrite the expression using the LCD. $\frac{3x}{x^4} + \frac{3 - x}{x^4}$

4. Add. $\frac{3x + (3 - x)}{x^4}$

5. Simplify the expression. $\frac{2x + 3}{x^4}$

Example 4 *Subtract with Unlike Denominators*

Simplify $\frac{1}{3n} - \frac{1 - 4n}{6n}$.

Solution

The LCD is $6n$.

$$\frac{1}{3n} - \frac{1 - 4n}{6n} = \frac{1}{3n} \cdot \frac{2}{2} - \frac{1 - 4n}{6n} \cdot \frac{1}{1}$$ **Rewrite using LCD.**

$$= \frac{2}{6n} - \frac{1 - 4n}{6n}$$ **Simplify numerators and denominators.**

$$= \frac{2 - (1 - 4n)}{6n}$$ **Subtract.**

$$= \frac{4n + 1}{6n}$$ **Simplify.**

Example 5 *Add with Unlike Binomial Denominators*

Simplify $\frac{x+3}{x-2} + \frac{8}{x+2}$.

Solution

Neither denominator can be factored. The least common denominator is the product ($x - 2$)($x + 2$) because it must contain both of these factors.

$\frac{x+3}{x-2} + \frac{8}{x+2}$ **Write original expression.**

$\frac{(x+3)(x+2)}{(x-2)(x+2)} + \frac{8(x-2)}{(x-2)(x+2)}$ **Rewrite using LCD.**

$\frac{x^2+5x+6}{(x-2)(x+2)} + \frac{8x-16}{(x-2)(x+2)}$ **Simplify numerators.**

$\frac{x^2+5x+6+(8x-16)}{(x-2)(x+2)}$ **Add.**

$\frac{x^2+13x-10}{(x-2)(x+2)}$ **Combine like terms.**

Checkpoint **Simplify the expression.**

1. $\frac{y-3}{2y^2} - \frac{9}{8y}$ $-\frac{5y+12}{8y^2}$	**2.** $\frac{x-1}{x+5} + \frac{3}{x-7}$ $\frac{x^2-5x+22}{(x+5)(x-7)}$

11.7 Rational Equations

Goal Solve rational equations.

VOCABULARY

Rational equation A rational equation is an equation that contains rational expressions.

Example 1 ***Cross Multiply***

Solve $\frac{4}{y+4} = \frac{y}{3}$.

1. **Write the original equation.** $\frac{4}{y+4} = \frac{y}{3}$
2. **Cross multiply.** $4(\underline{3}) = y(\underline{y+4})$
3. **Simplify each side.** $\underline{12} = \underline{y^2 + 4y}$
4. **Write the equation in standard form.** $0 = \underline{y^2 + 4y - 12}$
5. **Factor the right-hand side.** $0 = (\underline{y+6})(\underline{y-2})$

Answer The solutions are $\underline{-6}$ and $\underline{2}$.

When you solve rational equations, be sure to check your answers. Remember, values of the variable that make any denominator equal to 0 are excluded.

Example 2 ***Multiply by the LCD***

Solve $\frac{3}{x} + \frac{1}{4} = \frac{5}{2x}$.

1. **Find the LCD.** The LCD is $\underline{4x}$.
2. **Write the original equation.** $\frac{3}{x} + \frac{1}{4} = \frac{5}{2x}$
3. **Multiply each side by the LCD.** $\underline{4x} \cdot \frac{3}{x} + \underline{4x} \cdot \frac{1}{4} = \frac{5}{2x} \cdot \underline{4x}$
4. **Simplify each side.** $\underline{12 + x} = \underline{10}$
5. **Solve.** $x = \underline{-2}$

In Example 2, you must check that the solution does not result in a zero denominator in the original equation.

Example 3 *Factor First, then Multiply by the LCD*

Solve $\dfrac{1}{y-2} + \dfrac{2}{y^2 - 4y + 4} = 1.$

When a polynomial appears in the denominator of a fraction, you may want to factor the polynomial before solving the equation.

$$\frac{1}{y-2} + \frac{2}{y^2 - 4y + 4} = 1$$

$$\frac{1}{y-2} + \frac{2}{\boxed{(y-2)^2}} = 1$$

$$\frac{1}{y-2} \cdot \underline{(y-2)^2} + \frac{2}{\boxed{(y-2)^2}} \cdot \underline{(y-2)^2} = 1 \cdot \underline{(y-2)^2}$$

$$\underline{(y-2) + 2} = \underline{(y-2)^2}$$

$$\underline{y} = \underline{y^2 - 4y + 4}$$

$$0 = \underline{y^2 - 5y + 4}$$

$$0 = (\underline{y-1})(\underline{y-4})$$

Answer The solutions are $y = \underline{1}$ and $y = \underline{4}$. Check both values.

✔ ***Checkpoint*** **Solve the equation. Check your solutions.**

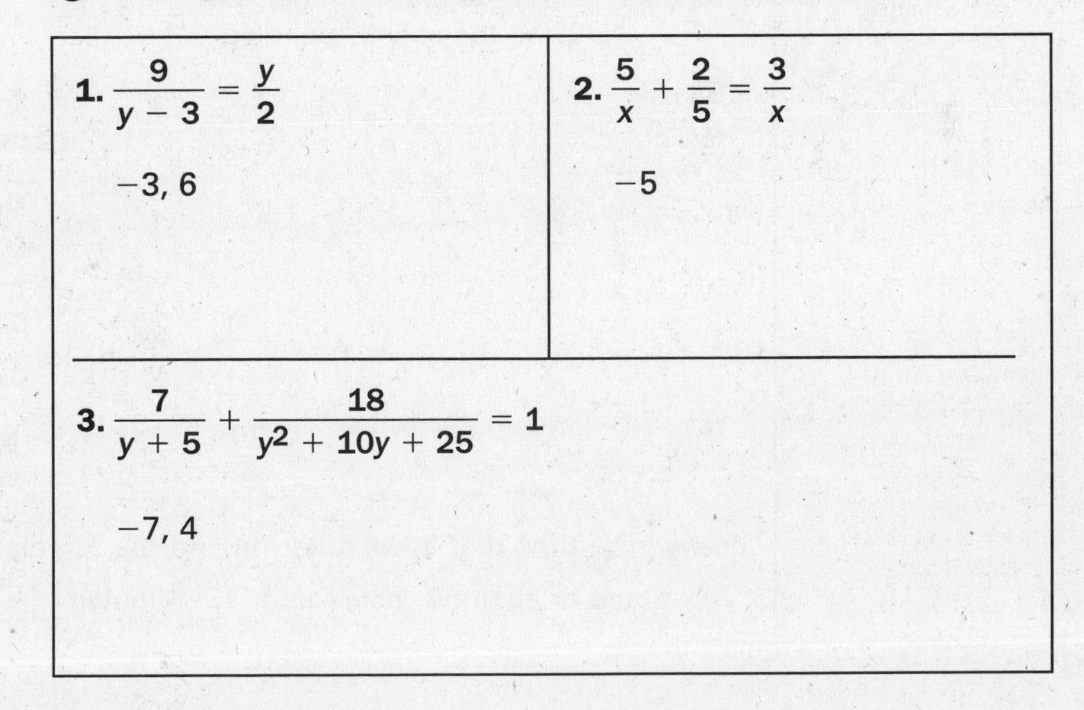

1. $\dfrac{9}{y-3} = \dfrac{y}{2}$

$-3, 6$

2. $\dfrac{5}{x} + \dfrac{2}{5} = \dfrac{3}{x}$

-5

3. $\dfrac{7}{y+5} + \dfrac{18}{y^2 + 10y + 25} = 1$

$-7, 4$

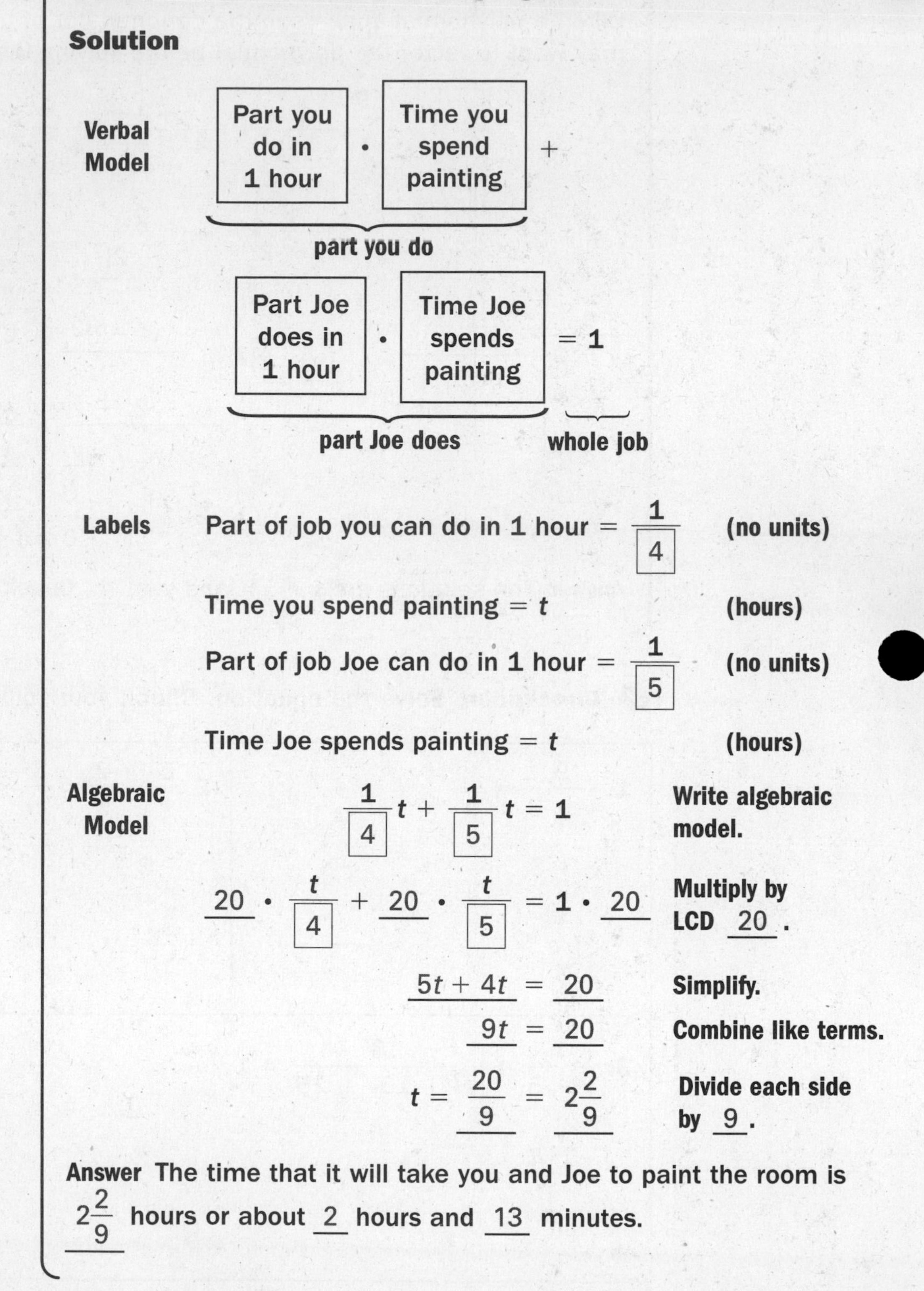

Example 4 ***Solve a Work Problem***

Painting Alone, you can paint your living room in 4 hours. Your friend Joe can paint the room in 5 hours. How long will it take you and Joe to paint the room, working together?

Solution

Verbal Model

[Part you do in 1 hour] • [Time you spend painting] + [Part Joe does in 1 hour] • [Time Joe spends painting] = 1

(part you do) (part Joe does) (whole job)

Labels

Part of job you can do in 1 hour $= \frac{1}{\boxed{4}}$ (no units)

Time you spend painting $= t$ (hours)

Part of job Joe can do in 1 hour $= \frac{1}{\boxed{5}}$ (no units)

Time Joe spends painting $= t$ (hours)

Algebraic Model

$$\frac{1}{\boxed{4}}t + \frac{1}{\boxed{5}}t = 1$$ **Write algebraic model.**

$$\underline{20} \cdot \frac{t}{\boxed{4}} + \underline{20} \cdot \frac{t}{\boxed{5}} = 1 \cdot \underline{20}$$ **Multiply by LCD $\underline{20}$.**

$$\underline{5t + 4t} = \underline{20}$$ **Simplify.**

$$\underline{9t} = \underline{20}$$ **Combine like terms.**

$$t = \underline{\frac{20}{9}} = \underline{2\frac{2}{9}}$$ **Divide each side by $\underline{9}$.**

Answer The time that it will take you and Joe to paint the room is $\underline{2\frac{2}{9}}$ hours or about $\underline{2}$ hours and $\underline{13}$ minutes.

Words to Review

Give an example of the vocabulary word.

Proportion	**Extremes**
$\frac{4}{y} = \frac{3}{7}$	In the proportion $\frac{4}{y} = \frac{3}{7}$, 4 and 7 are the extremes.
Means	**Inverse variation**
In the proportion $\frac{4}{y} = \frac{3}{7}$, y and 3 are the means.	$xy = 5$
Rational number	**Rational expression**
$\frac{1}{2}$	$\frac{2x}{x + 3}$
Least common denominator (LCD)	**Rational equation**
The least common denominator of $\frac{3}{4x}$ and $\frac{1}{6x^2}$ is $12x^2$.	$\frac{2x}{x + 3} = \frac{5}{x}$

Review your notes and Chapter 11 by using the Chapter Review on pages 681–684 of your textbook.

12.1 Functions Involving Square Roots

Goal Evaluate and graph a function involving square roots.

VOCABULARY

Square root function The square root function is defined by the equation $y = \sqrt{x}$.

Example 1 *Evaluate Functions Involving Square Roots*

Find the domain of $y = 3\sqrt{x}$. Use several values in the domain to make a table of values for the function.

Recall that the square root of a negative number is undefined. $\sqrt{x}$ can be evaluated only when $x \geq 0$.

Solution

A square root is defined only when the radicand is nonnegative.

Therefore the domain of $y = 3\sqrt{x}$ consists of all nonnegative real numbers. A table of values for $x = 0, 1, 2, 3$, and 4 is shown at the right.

x	y
0	$y = 3\sqrt{0} =$ 0
1	$y = 3\sqrt{1} =$ 3
2	$y = 3\sqrt{2} \approx$ 4.2
3	$y = 3\sqrt{3} \approx$ 5.2
4	$y = 3\sqrt{4} =$ 6

Example 2 *Graph $y = 3\sqrt{x}$*

Sketch the graph of $y = 3\sqrt{x}$. Then find its range.

Solution

From Example 1, you know the domain is all nonnegative real numbers. Use the table of values from Example 1. Then plot the points and connect them with a smooth curve. The range is all nonnegative real numbers.

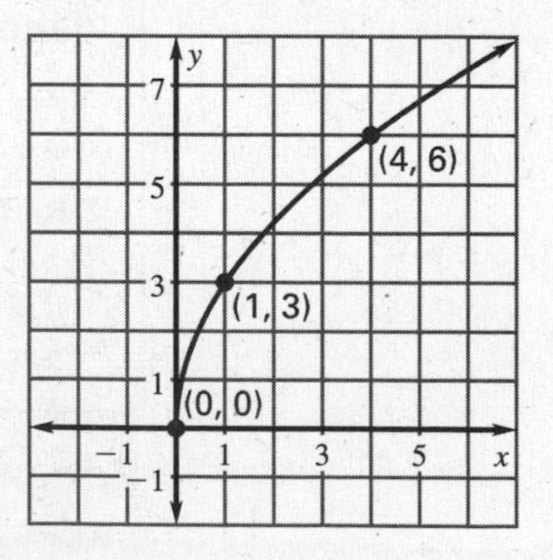

Example 3 *Graph $y = \sqrt{x} + 2$*

Find the domain of $y = \sqrt{x} + 2$. Then sketch its graph and find the range.

Solution

The domain is the values of x for which the radicand is nonnegative, so the domain consists of all nonnegative real numbers. Make a table of values, plot the points, and connect them with a smooth curve.

x	y
0	$y = \sqrt{0} + 2 = 2$
1	$y = \sqrt{1} + 2 = 3$
2	$y = \sqrt{2} + 2 \approx 3.4$
4	$y = \sqrt{4} + 2 = 4$
6	$y = \sqrt{6} + 2 \approx 4.4$

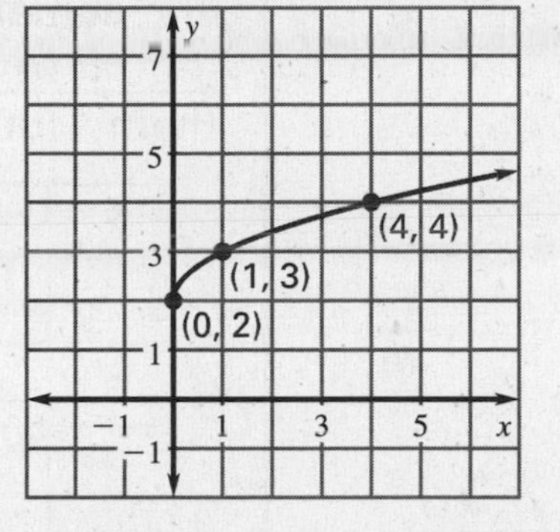

The range is all numbers that are greater than or equal to 2.

Checkpoint Find the domain of the function. Then sketch its graph and find the range.

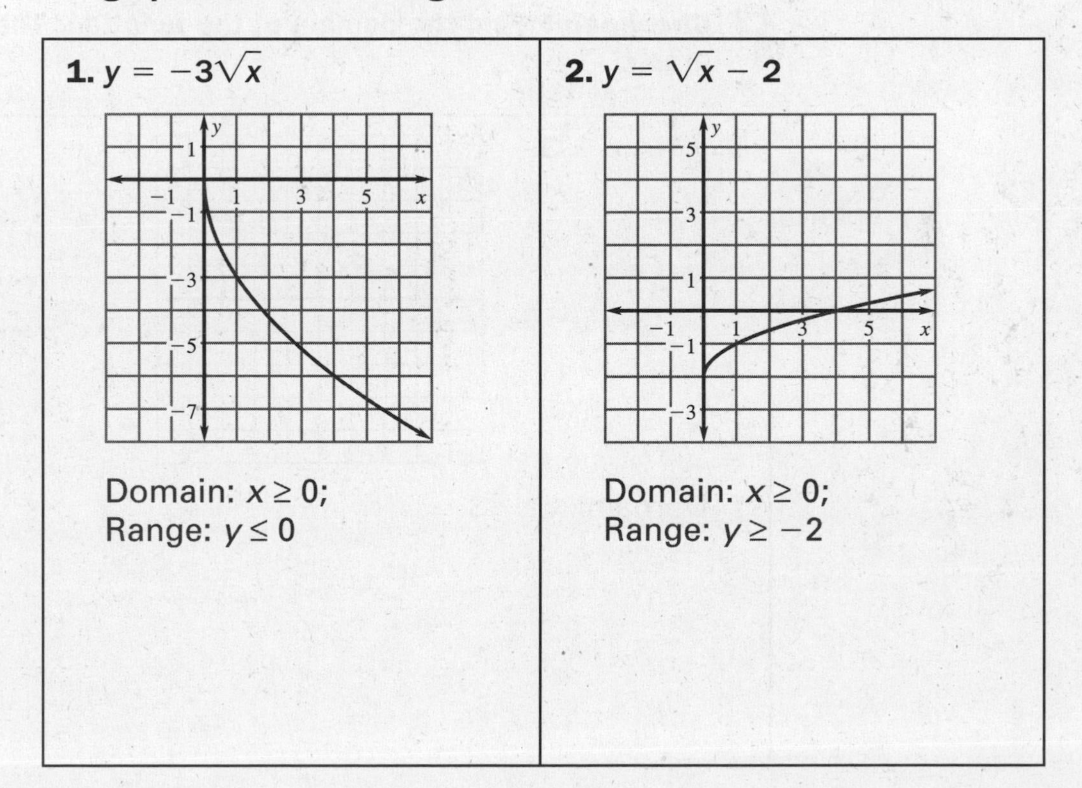

1. Domain: $x \geq 0$; Range: $y \leq 0$

2. Domain: $x \geq 0$; Range: $y \geq -2$

Example 4 *Graph $y = \sqrt{x - 2}$*

Find the domain of $y = \sqrt{x - 2}$. Then sketch its graph.

Solution

To find the domain, find the values of x for which the radicand is nonnegative.

$x - 2 \geq 0$ **Write an inequality for the domain.**

$x \geq 2$ **Add 2 to each side.**

The domain is the set of all numbers that are greater than or equal to 2. Make a table of values, plot the points, and connect them with a smooth curve.

x	y
2	$y = \sqrt{2 - 2} = 0$
3	$y = \sqrt{3 - 2} = 1$
4	$y = \sqrt{4 - 2} \approx 1.4$
6	$y = \sqrt{6 - 2} = 2$
8	$y = \sqrt{8 - 2} \approx 2.4$

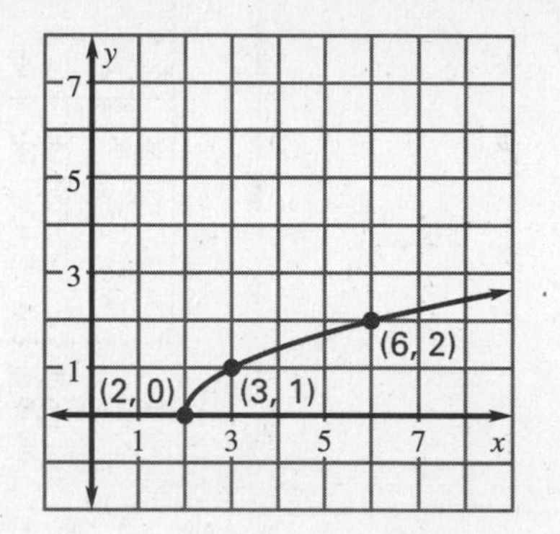

Checkpoint Find the domain of the function. Then sketch its graph.

3. $y = \sqrt{x + 3}$

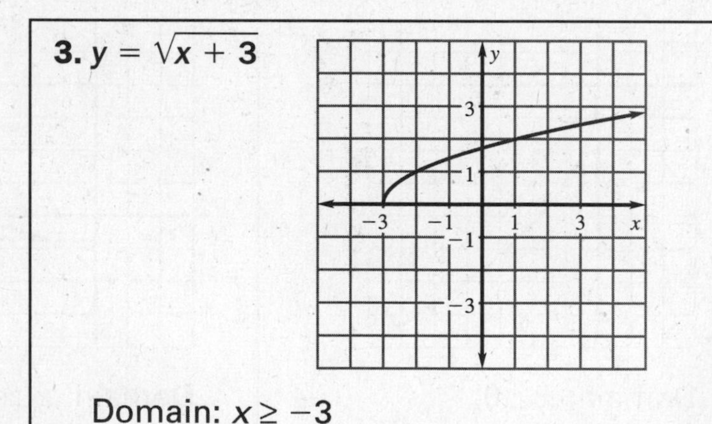

Domain: $x \geq -3$

12.2 Operations with Radical Expressions

Goal Add, subtract, multiply, and divide radical expressions.

Example 1 ***Add and Subtract Radicals***

Simplify the radical expression.

a. $4\sqrt{3} + 2\sqrt{7} - 9\sqrt{3}$

$= (\underline{4}\sqrt{3} - \underline{9}\sqrt{3}) + 2\sqrt{7}$ **Group radicals having the same radicand.**

$= \underline{-5}\sqrt{3} + 2\sqrt{7}$ **Subtract.**

b. $3\sqrt{5} - \sqrt{20} = 3\sqrt{5} - \sqrt{4 \cdot 5}$ **Factor using perfect square factor.**

$= 3\sqrt{5} - \underline{\sqrt{4}} \cdot \sqrt{5}$ **Use product property.**

$= 3\sqrt{5} - \underline{2\sqrt{5}}$ **Simplify.**

$= \underline{\sqrt{5}}$ **Subtract.**

✔ ***Checkpoint*** **Simplify the radical expression.**

1. $5\sqrt{6} + 4\sqrt{6}$ $9\sqrt{6}$	**2.** $4\sqrt{5} - 7\sqrt{5}$ $-3\sqrt{5}$
3. $6\sqrt{2} + \sqrt{18}$ $9\sqrt{2}$	**4.** $8\sqrt{7} - \sqrt{28}$ $6\sqrt{7}$

Example 2 *Multiply Radicals*

a. $\sqrt{4} \cdot \sqrt{9} = \underline{\sqrt{36}} = \underline{6}$ — Use product property and simplify.

b. $\sqrt{5}(3 - \sqrt{3}) = \underline{\sqrt{5}} \cdot 3 - \underline{\sqrt{5}} \cdot \sqrt{3}$ — Use distributive property.

$= \underline{3\sqrt{5} - \sqrt{15}}$ — Use product property.

c. $(1 + \sqrt{6})(1 - \sqrt{6}) = \underline{1}^2 - (\underline{\sqrt{6}})^2$ — Use sum and difference pattern.

$= \underline{1} - \underline{6}$ — Evaluate powers.

$= \underline{-5}$ — Simplify.

> Notice that the product of two radical expressions having the sum and difference pattern has no radical. In general, $(a + \sqrt{b})(a - \sqrt{b}) = a^2 - b$.

Example 3 *Simplify Radicals*

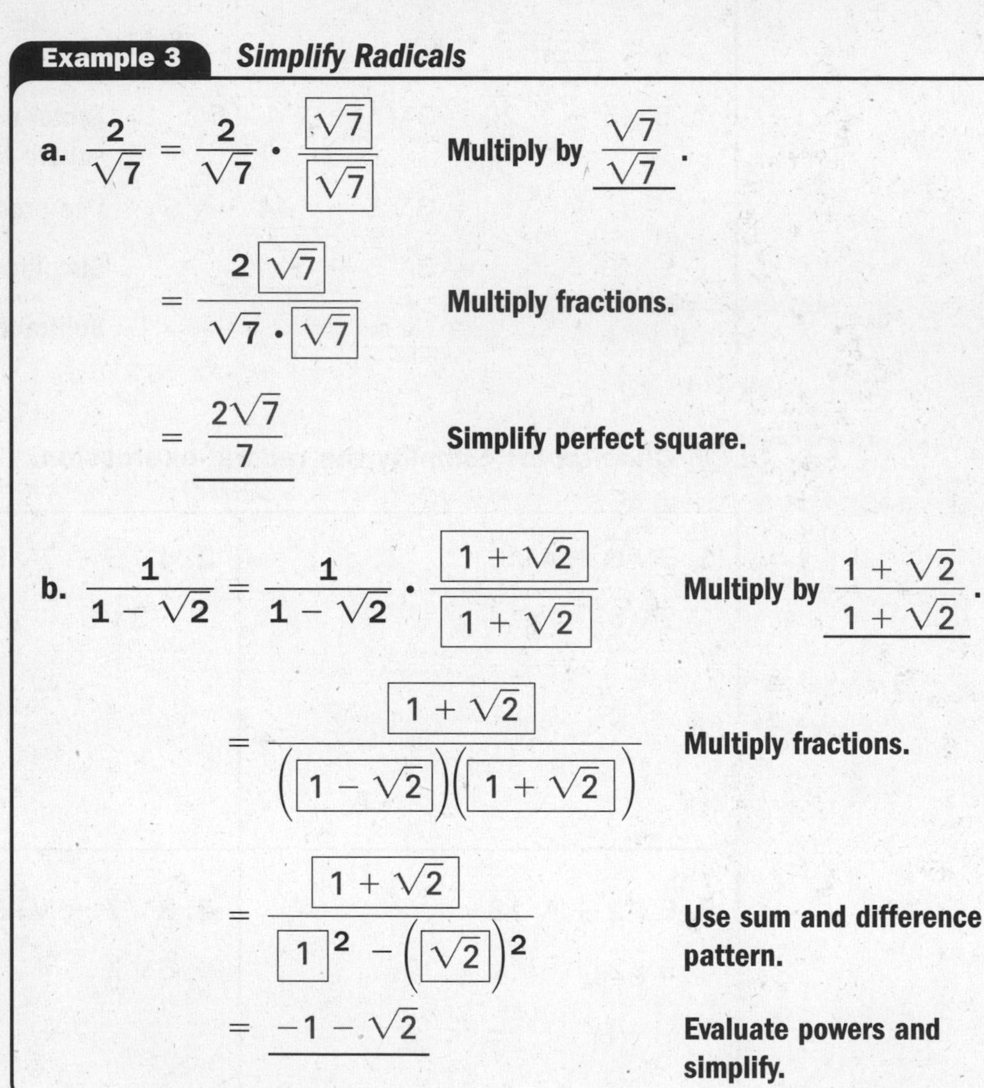

a. $\dfrac{2}{\sqrt{7}} = \dfrac{2}{\sqrt{7}} \cdot \dfrac{\boxed{\sqrt{7}}}{\boxed{\sqrt{7}}}$ — Multiply by $\underline{\dfrac{\sqrt{7}}{\sqrt{7}}}$.

$= \dfrac{2\boxed{\sqrt{7}}}{\sqrt{7} \cdot \boxed{\sqrt{7}}}$ — Multiply fractions.

$= \underline{\dfrac{2\sqrt{7}}{7}}$ — Simplify perfect square.

b. $\dfrac{1}{1 - \sqrt{2}} = \dfrac{1}{1 - \sqrt{2}} \cdot \dfrac{\boxed{1 + \sqrt{2}}}{\boxed{1 + \sqrt{2}}}$ — Multiply by $\underline{\dfrac{1 + \sqrt{2}}{1 + \sqrt{2}}}$.

$= \dfrac{\boxed{1 + \sqrt{2}}}{\left(\boxed{1 - \sqrt{2}}\right)\left(\boxed{1 + \sqrt{2}}\right)}$ — Multiply fractions.

$= \dfrac{\boxed{1 + \sqrt{2}}}{\boxed{1}^2 - \left(\boxed{\sqrt{2}}\right)^2}$ — Use sum and difference pattern.

$= \underline{-1 - \sqrt{2}}$ — Evaluate powers and simplify.

Checkpoint Simplify the radical expression.

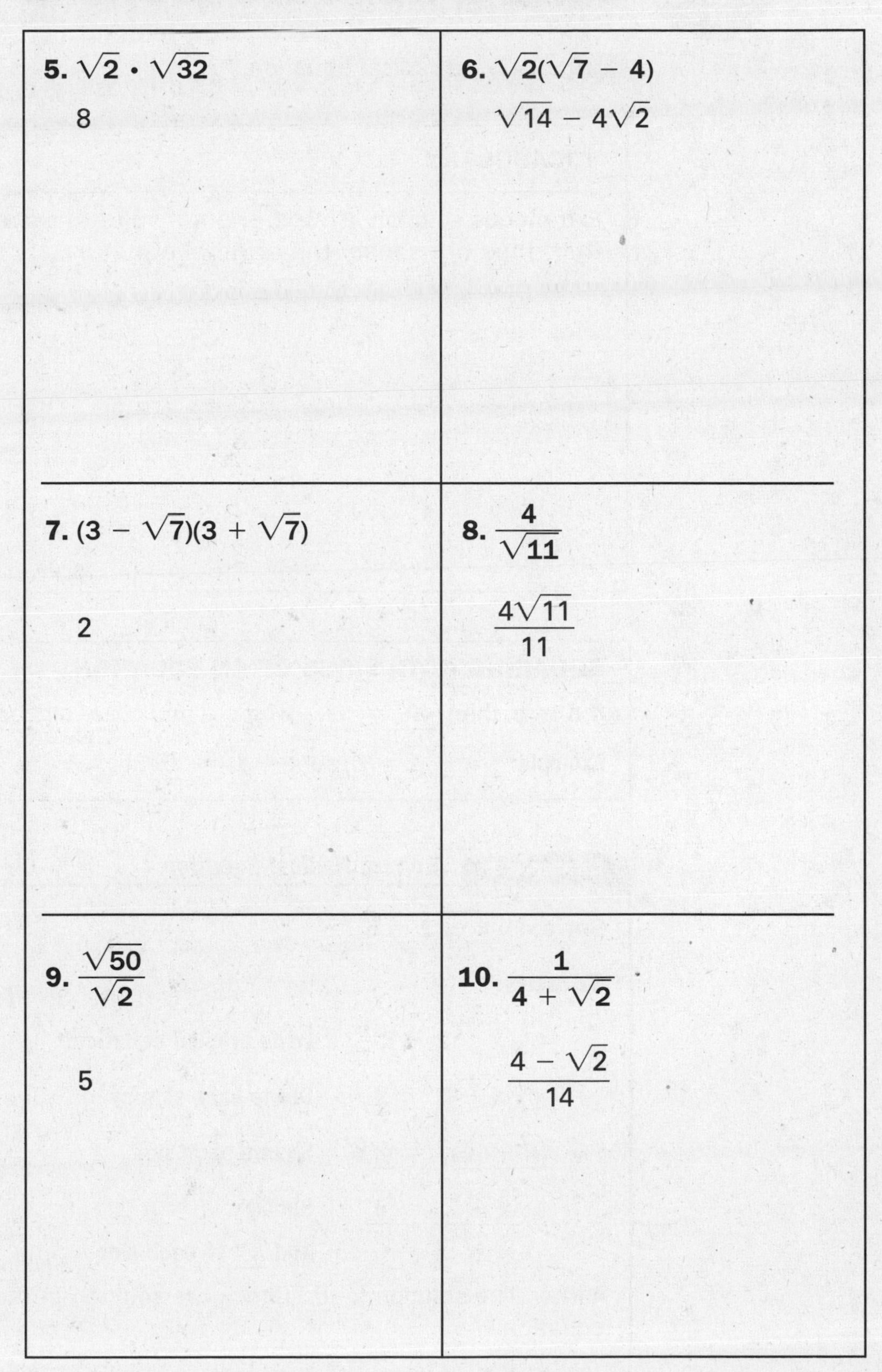

5. $\sqrt{2} \cdot \sqrt{32}$ 8	**6.** $\sqrt{2}(\sqrt{7} - 4)$ $\sqrt{14} - 4\sqrt{2}$
7. $(3 - \sqrt{7})(3 + \sqrt{7})$ 2	**8.** $\frac{4}{\sqrt{11}}$ $\frac{4\sqrt{11}}{11}$
9. $\frac{\sqrt{50}}{\sqrt{2}}$ 5	**10.** $\frac{1}{4 + \sqrt{2}}$ $\frac{4 - \sqrt{2}}{14}$

12.3 Solving Radical Equations

Goal Solve a radical equation.

VOCABULARY

Extraneous solution An extraneous solution is a trial solution that does not satisfy the original equation.

SQUARING BOTH SIDES OF AN EQUATION

If $a = b$, then $\underline{a^2} = \underline{b^2}$, where a and b are algebraic expressions.

Example: $\sqrt{x + 1} = 5$, so $x + 1 = \underline{25}$.

Example 1 ***Solve a Radical Equation***

Solve $4\sqrt{x - 2} = 8$

Solution

$4\sqrt{x - 2} = 8$	**Write original equation.**
$\sqrt{x - 2} = \underline{2}$	**Divide each side by** $\underline{4}$.
$(\underline{\sqrt{x - 2}})^2 = \underline{2}^2$	**Square each side.**
$\underline{x - 2} = \underline{4}$	**Simplify.**
$x = \underline{6}$	**Add** $\underline{2}$ **to each side.**

Answer The solution is $\underline{6}$. Check the solution in the original equation.

Example 2 *Solve a Radical Equation*

To solve the equation $\sqrt{3x - 6} + 4 = 7$, you need to isolate the radical expression first.

1. Write the original equation. $\sqrt{3x - 6} + 4 = 7$
2. Subtract 4 from each side. $\sqrt{3x - 6} = 3$
3. Square each side. $(\sqrt{3x - 6})^2 = 3^2$
4. Simplify the equation. $3x - 6 = 9$
5. Add 6 to each side. $3x = 15$
6. Divide each side by 3. $x = 5$

Answer The solution is 5. Check the solution in the original equation.

Example 3 *Check for Extraneous Solutions*

Solve $\sqrt{x + 6} = x$ and check for extraneous solutions.

1. Write the original equation. $\sqrt{x + 6} = x$
2. Square each side. $(\sqrt{x + 6})^2 = x^2$
3. Simplify the equation. $x + 6 = x^2$
4. Write equation in standard form. $0 = x^2 - x - 6$
5. Factor the equation. $0 = (x - 3)(x + 2)$
6. Use the zero-product property. $x = 3$ or $x = -2$

Check Substitute 3 and −2 in the original equation.

$\sqrt{3 + 6} \stackrel{?}{=} 3$ $\sqrt{-2 + 6} \stackrel{?}{=} -2$

$3 = 3$ $2 \neq -2$

Answer The only solution is 3, because $x = -2$ does not satisfy the original equation.

Example 4 *Check for Extraneous Solutions*

Solve $\sqrt{x} + 11 = 0$ and check for extraneous solutions.

Solution

$\sqrt{x} + 11 = 0$ — **Write original equation.**

$\sqrt{x} = \underline{-11}$ — **Subtract $\underline{11}$ from each side.**

$(\underline{\sqrt{x}})^2 = (\underline{-11})^2$ — **Square each side.**

$x = \underline{121}$ — **Simplify.**

Answer $\underline{\sqrt{121}} + 11 \neq 0$, so $x = \underline{121}$ is not a solution. The equation has no solution because $\sqrt{x} \geq \underline{0}$ for all values of x.

✔ ***Checkpoint*** **Solve the equation.**

1. $\sqrt{x} - 10 = 0$	**2.** $3\sqrt{x + 3} = 6$	**3.** $\sqrt{4x - 8} + 2 = 6$
100	1	6

Solve the equation. Check for extraneous solutions.

4. $\sqrt{2x + 3} = x$	**5.** $\sqrt{x} + 1 = 0$
3	no solution

12.4 Rational Exponents

Goal Evaluate expressions involving rational exponents.

VOCABULARY

Cube root If $b^3 = a$, then b is a cube root of a.

Rational exponent For any integer n and real number $a \geq 0$, the nth root of a is written $a^{1/n}$ or $\sqrt[n]{a}$. Let $a^{1/n}$ be an nth root of a, m be a positive integer, and $a \geq 0$.

Then $a^{m/n} = \left(a^{1/n}\right)^m = \left(\sqrt[n]{a}\right)^m = \sqrt[n]{a^m}$.

Example 1 ***Find Cube and Square Roots***

a. $8^{1/3}$ **b.** $\sqrt[3]{27}$ **c.** $81^{1/2}$

Solution

a. Because $\underline{2}^3 = 8$, you know that $8^{1/3} = \underline{2}$.

b. Because $\underline{3}^3 = 27$, you know that $\sqrt[3]{27} = \underline{3}$.

c. Because $\underline{9}^2 = 81$, and $\underline{9}$ is a nonnegative number, you know that $81^{1/2} = \underline{9}$.

RATIONAL EXPONENTS

Let a be a nonnegative number, and let m and n be positive integers.

$a^{m/n} = \left(a^{1/n}\right)^m = \left(\sqrt[n]{a}\right)^m$

Example 2 *Evaluate Expressions with Rational Exponents*

Rewrite the expression using rational exponent notation *and* radical notation.

a. $9^{3/2}$

b. $16^{5/4}$

Solution

a. Use rational exponent notation. $9^{3/2} = \left(9^{\underline{1/2}}\right)^{\underline{3}} = \underline{3^3} = \underline{27}$

Use radical notation. $9^{3/2} = \left(\underline{\sqrt{9}}\right)^{\underline{3}} = \underline{3^3} = \underline{27}$

b. Use rational exponent notation.

$16^{5/4} = \left(16^{\underline{1/4}}\right)^{\underline{5}} = \underline{2^5} = \underline{32}$

Use radical notation. $16^{5/4} = \left(\sqrt[\boxed{4}]{16}\right)^{\underline{5}} = \underline{2^5} = \underline{32}$

Checkpoint Evaluate the expression without using a calculator.

1. $144^{1/2}$	2. $\sqrt[3]{125}$	3. $4^{5/2}$	4. $(\sqrt{16})^3$
12	5	32	64

PROPERTIES OF RATIONAL EXPONENTS

Let *a* and *b* be nonnegative real numbers and let *m* and *n* be rational numbers.

Property	Example
$a^m \cdot a^n = a^{m+n}$	$3^{1/2} \cdot 3^{3/2} = 3^{(1/2 + 3/2)} = 3^{\underline{2}} = \underline{9}$
$(a^m)^n = a^{mn}$	$(4^{3/2})^2 = 4^{3/2 \cdot 2} = 4^{\underline{3}} = \underline{64}$
$(ab)^m = a^m b^m$	$(9 \cdot 4)^{1/2} = 9^{1/2} \cdot 4^{1/2} = \underline{3} \cdot \underline{2} = \underline{6}$

Example 3 ***Use Properties of Rational Exponents***

Evaluate the expression using the properties of rational exponents.

a. $2^{1/4} \cdot 2^{3/4}$ **b.** $(2^{1/3})^9$ **c.** $(9 \cdot 16)^{1/2}$

Solution

a. Use the product of powers property.

$$2^{1/4} \cdot 2^{3/4} = 2^{(1/4\ +\ 3/4)} = 2^{4/4} = 2^{1} = 2$$

b. Use the power of a power property.

$$(2^{1/3})^9 = 2^{(1/3\ \cdot\ 9)} = 2^{3} = 8$$

c. Use the power of a product property.

$$(9 \cdot 16)^{1/2} = 9^{1/2} \cdot 16^{1/2} = 3 \cdot 4 = 12$$

Example 4 ***Use Properties of Rational Exponents***

Simplify the variable expression $(a^{1/3} \cdot b)^3 \sqrt{a}$ using the properties of rational exponents.

1. Use power of a product property. $(a^{1/3} \cdot b)^3 \sqrt{a}$

$= (a^{(1/3\ \cdot\ 3)} \cdot b^{3})\sqrt{a}$

2. Write $\sqrt{a}$ in rational exponent notation. $= a^{1} \cdot b^{3} \cdot a^{1/2}$

3. Use product of powers property. $= a^{1\ +\ 1/2} \cdot b^{3}$

4. Add the exponents. $= a^{3/2} b^{3}$

✔ ***Checkpoint*** **Evaluate the expression using the properties of rational exponents.**

5. $3^{5/2} \cdot 3^{3/2}$	**6.** $(8 \cdot 27)^{1/3}$
81	6

12.5 Completing the Square

Goal Solve a quadratic equation by completing the square.

VOCABULARY

Completing the square Completing the square is the process of rewriting a quadratic formula so that one side is a perfect square trinomial.

COMPLETING THE SQUARE

To complete the square of the expression $x^2 + bx$, add the square of half the coefficient of x, that is, add $\left(\frac{b}{2}\right)^2$.

$$x^2 + bx + \left(\frac{b}{2}\right)^2 = \left(x + \frac{b}{2}\right)^2$$

Example 1 ***Complete the Square***

What term should you add to $x^2 + 12x$ to create a perfect square trinomial?

Solution

The coefficient of x is 12, so you should add $\left(\frac{12}{2}\right)^2$, or 36, to the expression.

$$x^2 + 12x + \left(\frac{12}{2}\right)^2 = x^2 + 12x + 36$$

$$= (x + 6)^2$$

Example 2 *Solve a Quadratic Equation*

Solve $x^2 + 6x = 16$ by completing the square.

Solution

$x^2 + 6x = 16$	**Write original equation.**
$x^2 + 6x + \underline{9} = 16 + \underline{9}$	**Add,** $\left(\frac{\boxed{6}}{2}\right)^2$**, or** $\underline{9}$ **, to each side.**
$(x + \underline{3})^2 = \underline{25}$	**Write left side as perfect square.**
$x + \underline{3} = \underline{\pm 5}$	**Find square root of each side.**
$x = \underline{-3} \pm \underline{5}$	**Subtract** $\underline{3}$ **from each side.**
$x = \underline{2}$ or $x = \underline{-8}$	**Simplify.**

Answer The solutions are $\underline{2}$ and $\underline{-8}$. Check these in the original equation to confirm that both are solutions.

✓ *Checkpoint* **Solve the equation by completing the square.**

1. $x^2 + 8x = 48$ $-12, 4$	**2.** $x^2 - 4x - 4 = 0$ $2 \pm 2\sqrt{2}$

METHODS FOR SOLVING QUADRATIC EQUATIONS

Method	Comments
Finding Square Roots *(Lesson 9.2)*	Efficient way to solve $ax^2 + c = 0$.
Graphing *(Lesson 9.5)*	Can be used for *any* quadratic equation. Enables you to approximate solutions.
Using the Quadratic Formula *(Lesson 9.6)*	Can be used for *any* quadratic equation.
Factoring *(Lessons 10.5–10.8)*	Efficient way to solve a quadratic equation if the quadratic expression can be factored easily.
Completing the Square *(Lesson 12.5)*	Can be used for *any* quadratic equation, but is best suited for quadratic equations where $a = 1$ and b is an even number.

Example 3 ***Choose a Solution Method***

Choose a method and solve $5x^2 - 20 = 0$.

Solution

Because this quadratic equation is of the form $ax^2 + c = 0$, it is most efficiently solved by finding square roots.

$5x^2 - 20 = 0$	**Write original equation.**
$5x^2 = 20$	**Add 20 to each side.**
$x^2 = 4$	**Divide each side by 5.**
$x = \pm 2$	**Find square root of each side.**

Answer The solutions are 2 and -2. Check both solutions in the original equation.

12.6 The Pythagorean Theorem and Its Converse

Goal Use the Pythagorean theorem and its converse.

VOCABULARY

Theorem A theorem is a statement that has been proven to be true.

Pythagorean theorem The Pythagorean theorem is a theorem that states a relationship among the sides of a right triangle.

Hypotenuse The hypotenuse is the side opposite the right angle in a right triangle.

Legs of a right triangle The legs of a right triangle are the two sides of a right triangle that are not opposite the right angle.

Converse The converse of the statement "If *p*, then *q*" is the related statement "If *q*, then *p*," in which the hypothesis and conclusion are interchanged.

THE PYTHAGOREAN THEOREM

If a triangle is a right triangle, then the sum of the squares of the lengths of the legs *a* and *b* equals the square of the length of the hypotenuse *c*.

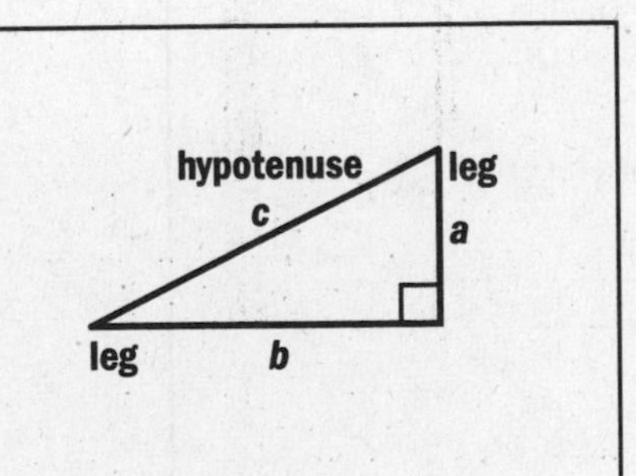

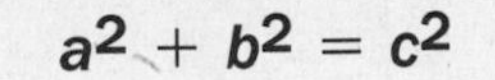

$$a^2 + b^2 = c^2$$

Example 1 *Use the Pythagorean Theorem*

When you use the Pythagorean theorem to find the length of a side of a right triangle, you need only the positive square root because the length of a side cannot be negative.

a. Given $a = 5$ and $b = 12$, find c. Use the Pythagorean theorem: $a^2 + b^2 = c^2$.

$$\underline{5^2 + 12^2} = c^2$$
$$\underline{169} = c^2$$
$$\underline{\sqrt{169}} = c$$
$$\underline{13} = c$$

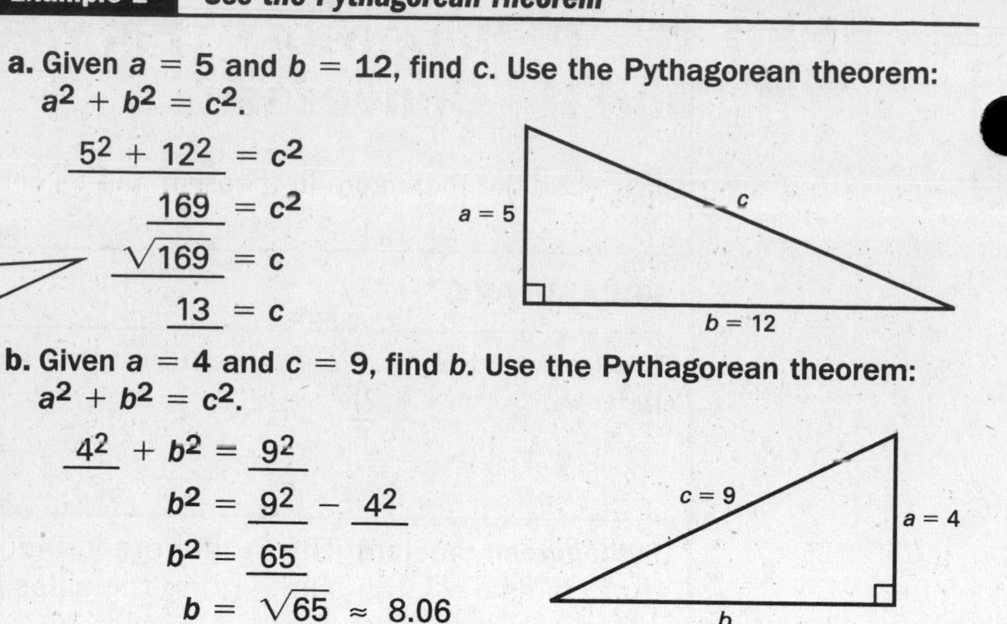

b. Given $a = 4$ and $c = 9$, find b. Use the Pythagorean theorem: $a^2 + b^2 = c^2$.

$$\underline{4^2} + b^2 = \underline{9^2}$$
$$b^2 = \underline{9^2} - \underline{4^2}$$
$$b^2 = \underline{65}$$
$$b = \underline{\sqrt{65}} \approx \underline{8.06}$$

✔ ***Checkpoint*** **Find the missing length of the right triangle where *a* and *b* are the lengths of the legs and *c* is the length of the hypotenuse.**

1. $a = 8$, $b = 15$	**2.** $a = 10$, $c = 20$
17	$10\sqrt{3}$

Example 2 *Use the Pythagorean Theorem*

A right triangle has one leg that is 2 inches longer than the other leg. The hypotenuse is 10 inches. Find the unknown lengths.

Solution

Sketch a right triangle and label the sides. Let x be the length of the shorter leg. Use the Pythagorean theorem to solve for x.

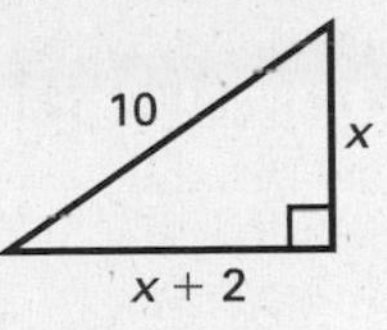

$a^2 + b^2 = c^2$	**Write Pythagorean theorem.**
$x^2 + (x + 2)^2 = 10^2$	**Substitute for *a*, *b*, and *c*.**
$x^2 + x^2 + 4x + 4 = 100$	**Simplify.**
$2x^2 + 4x - 96 = 0$	**Write in standard form.**
$2(x - 6)(x + 8) = 0$	**Factor.**
$x = 6$ *or* $x = -8$	**Zero-product property**

Answer Length is positive, so the solution $x = -8$ is extraneous. The sides have lengths 6 inches and $6 + 2 = 8$ inches.

CONVERSE OF THE PYTHAGOREAN THEOREM

If a triangle has side lengths a, b, and c such that $a^2 + b^2 = c^2$, then the triangle is a right triangle.

Example 3 *Determine Right Triangles*

Determine whether the given lengths are sides of a right triangle: 6, 8, 11.

In a right triangle, the hypotenuse is always the longest side.

Solution

Use the converse of the Pythagorean theorem. The lengths are not sides of a right triangle because

$6^2 + 8^2 = 36 + 64 = 100 \neq 11^2$.

12.7 The Distance Formula

Goal Find the distance between two points in a coordinate plane.

THE DISTANCE FORMULA

The distance d between the points (x_1, y_1) and (x_2, y_2) is

$$d = \sqrt{(x_2 - x_1)^2 + (y_2 - y_1)^2}.$$

Example 1 ***Find the Distance Between Two Points***

Use the distance formula to find the distance between (2, 4) and (4, −3).

$d = \sqrt{(x_2 - x_1)^2 + (y_2 - y_1)^2}$ **Write distance formula.**

$= \sqrt{(4 - 2)^2 + (-3 - 4)^2}$ **Substitute.**

$= \sqrt{(2)^2 + (-7)^2}$ **Simplify.**

$= \sqrt{4 + 49}$ **Evaluate powers.**

$= \sqrt{53}$ **Add.**

≈ 7.28 **Use a calculator.**

✔ ***Checkpoint*** **Find the distance between the points. Round your solution to the nearest hundredth if necessary.**

1. (3, 2), (−5, 2)	**2.** (−4, 5), (6, −2)
8	12.21

Example 2 *Check a Right Triangle*

Vertices is the plural form of *vertex*. A triangle has three vertices.

Determine whether the points (1, 0), (3, −1), and (4, 6) are vertices of a right triangle.

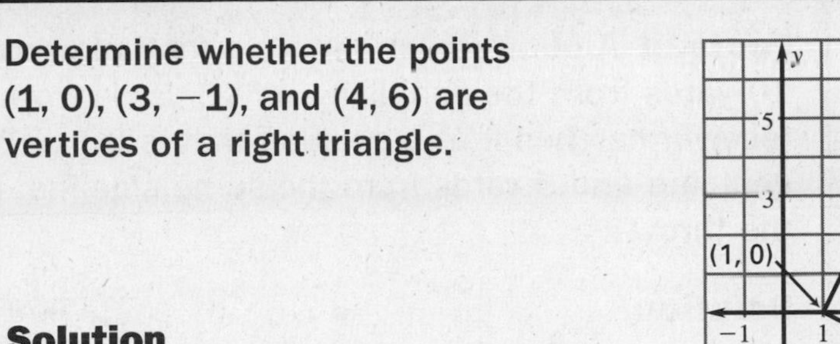

Solution

Use the distance formula to find the lengths of the three sides.

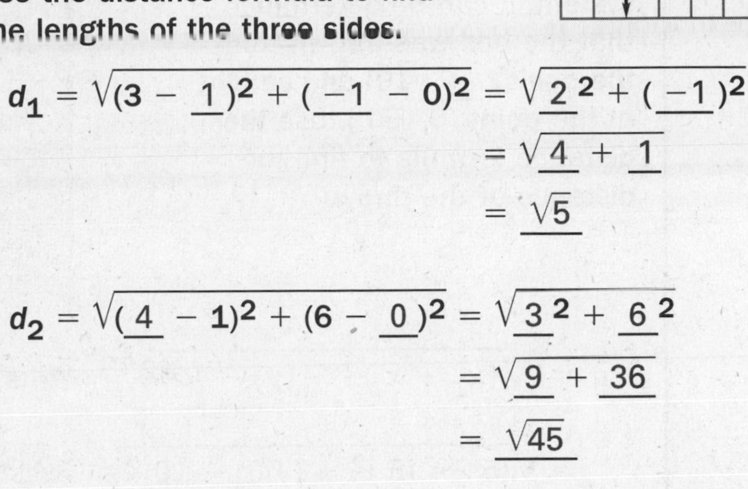

$$d_1 = \sqrt{(3 - \underline{1})^2 + (\underline{-1} - 0)^2} = \sqrt{\underline{2}^2 + (\underline{-1})^2}$$

$$= \sqrt{\underline{4} + \underline{1}}$$

$$= \underline{\sqrt{5}}$$

$$d_2 = \sqrt{(\underline{4} - 1)^2 + (6 - \underline{0})^2} = \sqrt{\underline{3}^2 + \underline{6}^2}$$

$$= \sqrt{\underline{9} + \underline{36}}$$

$$= \underline{\sqrt{45}}$$

$$d_3 = \sqrt{(4 - \underline{3})^2 + [6 - (\underline{-1})]^2} = \sqrt{\underline{1}^2 + \underline{7}^2}$$

$$= \sqrt{\underline{1} + \underline{49}}$$

$$= \underline{\sqrt{50}}$$

Next find the sum of the squares of the lengths of the two shorter sides.

$d_1^2 + d_2^2 = \underline{(\sqrt{5})^2} + \underline{(\sqrt{45})^2}$	**Substitute for d_1 and d_2.**
$= \underline{5 + 45}$	**Simplify.**
$= \underline{50}$	**Add.**

The sum of the squares of the lengths of the two shorter sides is $\underline{50}$, which $\underline{\text{is equal to}}$ the square of the length of the longest side, $\underline{(\sqrt{50})^2}$.

Answer By the converse of the Pythagorean theorem, the given points $\underline{\text{are}}$ vertices of a right triangle.

Example 3 *Apply the Distance Formula*

Football A quarterback throws a football from a position that is 10 yards from the goal line and 15 yards from the sideline. His receiver catches it at a position that is 50 yards from the same goal line and 5 yards from the same sideline. Find the distance of the throw.

Solution

By applying a coordinate system, it can be determined that the ball was thrown from the point (15, 10) and caught at the point (5, 50). Use the distance formula to find the distance of the throw.

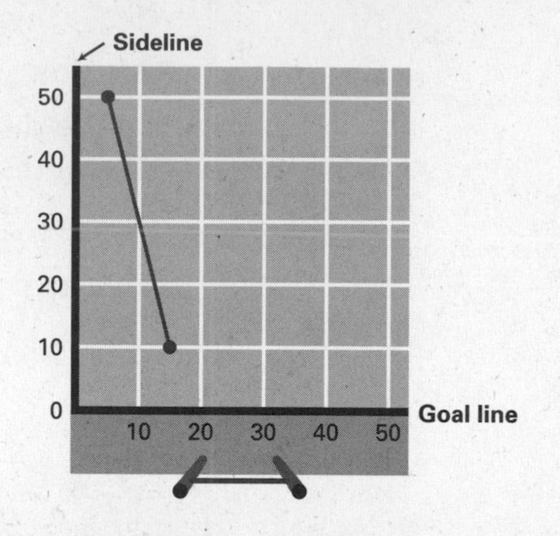

$d = \sqrt{(x_2 - x_1)^2 + (y_2 - y_1)^2}$	**Write the distance formula.**
$= \sqrt{(5 - 15)^2 + (50 - 10)^2}$	**Substitute.**
$= \sqrt{(-10)^2 + (40)^2}$	**Simplify.**
$= \sqrt{100 + 1600}$	**Evaluate powers.**
$= \sqrt{1700}$	**Add.**
≈ 41.2	**Use a calculator.**

Answer The ball was thrown about 41 yards.

Checkpoint **Complete the following exercise.**

3. Determine whether the points (0, 4), (2, −2), and (5, −1) are the vertices of a right triangle.

yes

12.8 The Midpoint Formula

Goal Find the midpoint of a line segment in a coordinate plane.

VOCABULARY

Midpoint The midpoint of a line segment is the point on the segment that is equidistant from its endpoints.

THE MIDPOINT FORMULA

The midpoint between (x_1, y_1) and (x_2, y_2) is

$$\left(\frac{x_1 + x_2}{2}, \frac{y_1 + y_2}{2}\right).$$

Example 1 ***Find the Midpoint***

Find the midpoint of the line segment connecting the points $(-3, -1)$ and $(3, 4)$. Use a graph to explain the result.

Solution

Let $(-3, -1) = (x_1, y_1)$ and $(3, 4) = (x_2, y_2)$.

$$\left(\frac{x_1 + x_2}{2}, \frac{y_1 + y_2}{2}\right) = \left(\frac{\boxed{-3} + \boxed{3}}{2}, \frac{\boxed{-1} + \boxed{4}}{2}\right) = \left(\frac{\boxed{0}}{2}, \frac{\boxed{3}}{2}\right)$$

Answer The midpoint is $\left(0, \frac{3}{2}\right)$.

From the graph, you can see that the point $\left(0, \frac{3}{2}\right)$ appears to be halfway between $(-3, -1)$ and $(3, 4)$.

In Example 2 you will use the distance formula to check a midpoint.

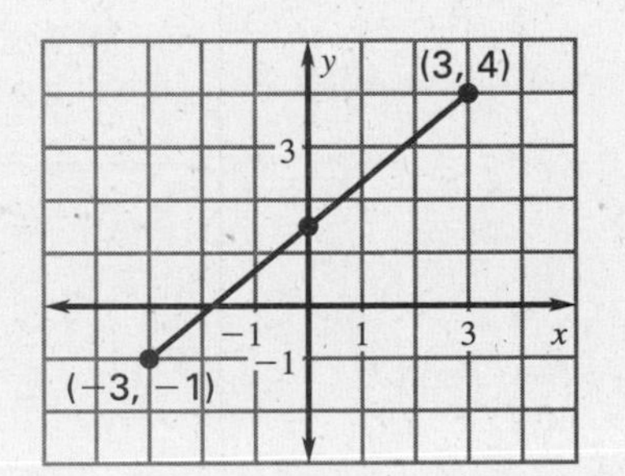

Example 2 *Check a Midpoint*

Use the distance formula to check the midpoint in Example 1.

Solution

The distance between $\left(0, \frac{3}{2}\right)$ and $(-3, -1)$ is:

$$\sqrt{(-3 - \underline{0})^2 + \left(-1 - \underline{\frac{3}{2}}\right)^2} = \sqrt{(\underline{-3})^2 + \left(\underline{-\frac{5}{2}}\right)^2}$$

$$= \sqrt{\underline{9} + \underline{\frac{25}{4}}} = \sqrt{\underline{\frac{61}{4}}} = \underline{\frac{\sqrt{61}}{2}}.$$

The distance between $\left(0, \frac{3}{2}\right)$ and $(3, 4)$ is:

$$\sqrt{(3 - \underline{0})^2 + \left(4 - \underline{\frac{3}{2}}\right)^2} = \sqrt{\underline{3}^2 + \left(\underline{\frac{5}{2}}\right)^2}$$

$$= \sqrt{\underline{9} + \underline{\frac{25}{4}}} = \sqrt{\underline{\frac{61}{4}}} = \underline{\frac{\sqrt{61}}{2}}.$$

Answer The distances from $\left(0, \frac{3}{2}\right)$ to the ends of the segment are equal.

✔ ***Checkpoint*** **Find the midpoint of the line segment connecting the given points.**

1. $(3, 5), (-2, -3)$ $\left(\frac{1}{2}, 1\right)$	**2.** $(-7, 4), (5, -10)$ $(-1, -3)$
3. $(-9, -2), (6, -2)$ $\left(-\frac{3}{2}, -2\right)$	**4.** $(-6, 9), (2, -5)$ $(-2, 2)$

Logical Reasoning: Proof

Goal Use logical reasoning and proof to prove that a statement is true or false.

VOCABULARY

Postulates or axioms Postulates (or axioms) are basic properties of mathematics that mathematicians accept without proof.

Conjecture A conjecture is a statement that is thought to be true but is not yet proved. Often it is a statement based on observation.

Indirect proof An indirect proof is a type of proof where it is first assumed that the statement is false. If this assumption leads to an impossibility, then the original statement has been proved to be true.

THE BASIC AXIOMS OF ALGEBRA

Let a, b, and c be real numbers.

Axioms of Addition and Multiplication

Closure: $a + b$ is a real number
ab is a real number.

Commutative: $a + b = b + a$
$ab = ba$

Associative: $(a + b) + c = a + (b + c)$
$(ab)c = a(bc)$

Identity: $a + 0 = a$, $0 + a = a$
$a(1) = a$, $1(a) = a$

Inverse: $a + (-a) = 0$
$a\left(\frac{1}{a}\right) = 1$, $a \neq 0$

THE BASIC AXIOMS OF ALGEBRA (CONT.)

Axiom Relating Addition and Multiplication

Distributive: $a(b + c) = \underline{ab} + \underline{ac}$

$(a + b)c = \underline{ac} + \underline{bc}$

Axioms of Equality

Addition: If $a = b$, then $a + \underline{c} = b + \underline{c}$.

Multiplication: If $a = b$, then $a\underline{c} = b\underline{c}$.

Substitution: If $a = b$, then a can be substituted for $\underline{b}$.

When you are proving a theorem, every step must be justified by an axiom, a definition, given information, or a previously proved theorem.

Example 1 *Prove a Theorem*

Use properties of algebra to prove the following theorem:

If a, b, and c are real numbers, then $(a + b) - c = (b - c) + a$.

Solution

$(a + b) - c = \underline{a} + (\underline{b} - \underline{c})$ Associative **property**

$= (\underline{b} - \underline{c}) + \underline{a}$ Commutative **property**

Example 2 *Find a Counterexample*

Show that the statement below is false by finding a counterexample.

For all numbers a and b, $a^2 + b^2 = (a + b)^2$.

Solution

The statement claims that $a^2 + b^2 = (a + b)^2$ for all values of a and b. If we let $a = 2$ and $b = 3$, we find $a^2 + b^2 = \underline{2}^2 + \underline{3}^2 = \underline{13}$, but $(a + b)^2 = (\underline{2} + \underline{3})^2 = \underline{25}$. Since $\underline{13} \neq \underline{25}$, the counterexample $a = 2$ and $b = 3$ shows that the general statement proposed above is false.

 ***Checkpoint*** **Complete the following exercise.**

1. Assign values to *a* and *b* to show that the following conjecture is false.

$-(a + b) = -a + b$

Sample Answer: $a = 5$, $b = 7$

Example 3 ***Use an Indirect Proof***

Use an indirect proof to prove that $\sqrt{3}$ is an irrational number.

Solution

If you assume that $\sqrt{3}$ is __not__ an irrational number, then $\sqrt{3}$ is __rational__ and can be written as the quotient of two integers *a* and *b* that have no common factors other than 1.

$\sqrt{3} = \frac{a}{b}$ **Assume $\sqrt{3}$ is a rational number.**

$3 = \frac{a^2}{b^2}$ **Square each side.**

$3\underline{b^2} = a^2$ **Multiply each side by $\underline{b^2}$.**

This implies that 3 is a factor of $\underline{a^2}$. Therefore 3 is also a factor of $\underline{a}$. Thus *a* can be written as $3c$.

$3b^2 = (3c)^2$ **Substitute $3c$ for a.**

$3b^2 = \underline{9}c^2$ **Simplify.**

$b^2 = \underline{3}c^2$ **Divide each side by 3.**

This implies that __3__ is a factor of b^2 and also a factor of *b*. So, __3__ is a factor of both *a* and *b*. But this is __impossible__ because *a* and *b* have no common factors other than 1. Therefore it is __impossible__ that $\sqrt{3}$ is a rational number. So you can conclude that $\sqrt{3}$ __must be__ an irrational number.

Words to Review

Give an example of the vocabulary word.

Square root function $y = \sqrt{x}$	**Extraneous solution** $x = 64$ is an extraneous solution of $\sqrt{x} + 8 = 0$.
Cube root of *a* 3 is a cube root of 27.	**Rational exponent** $\frac{1}{3}$ is the rational exponent in $a^{1/3}$.
Completing the square To complete the square of $x^2 + 6x$, add 9.	**Theorem** Pythagorean theorem $a^2 + b^2 = c^2$
Pythagorean theorem For a right triangle with legs a and b and hypotenuse c, $a^2 + b^2 = c^2$.	**Hypotenuse** c a b In the triangle above, side c is the hypotenuse.

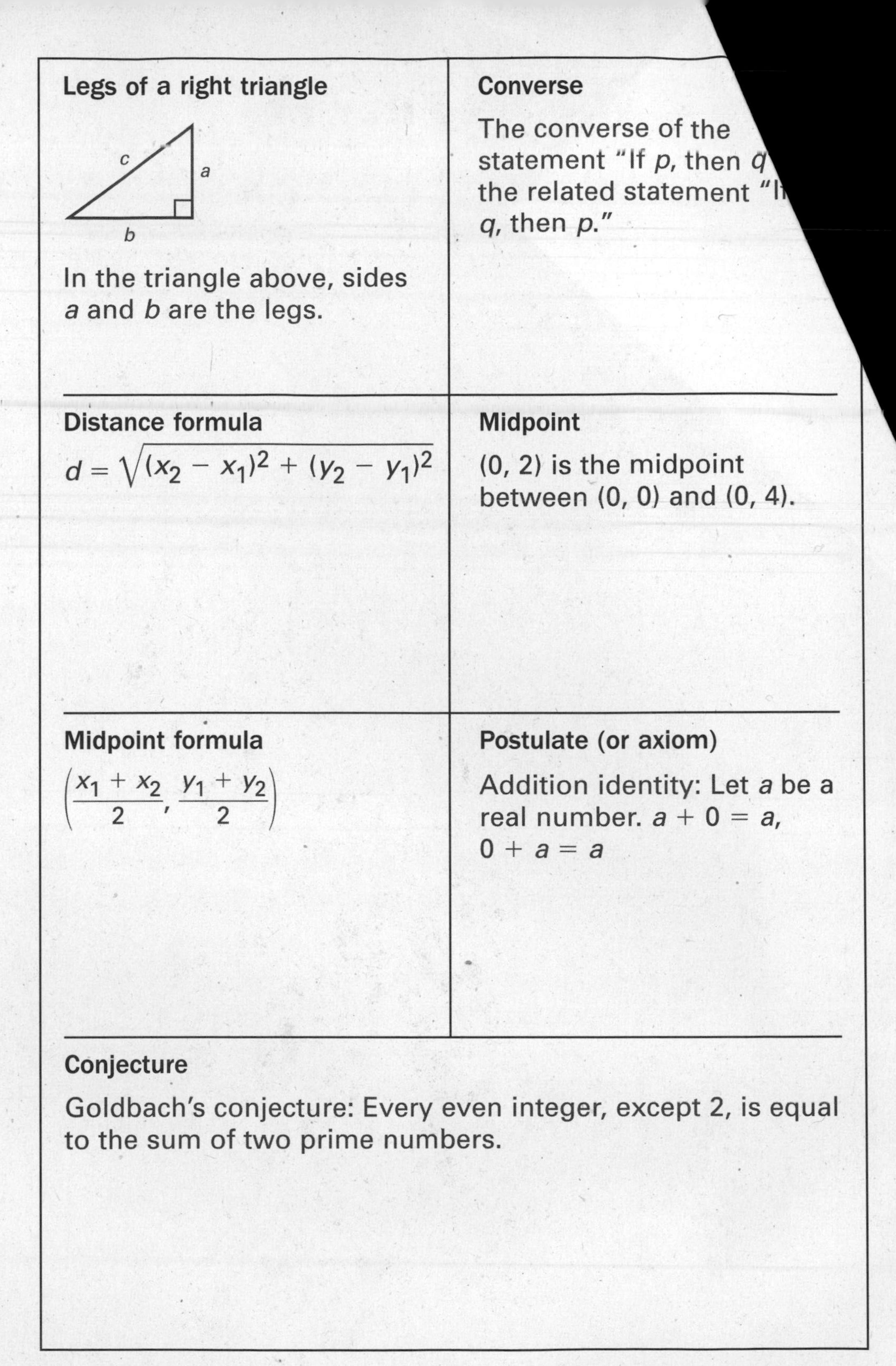

Legs of a right triangle (triangle with sides labelled c, a, b) In the triangle above, sides *a* and *b* are the legs.	**Converse** The converse of the statement "If *p*, then *q* the related statement "I *q*, then *p*."
Distance formula $d = \sqrt{(x_2 - x_1)^2 + (y_2 - y_1)^2}$	**Midpoint** (0, 2) is the midpoint between (0, 0) and (0, 4).
Midpoint formula $\left(\frac{x_1 + x_2}{2}, \frac{y_1 + y_2}{2}\right)$	**Postulate (or axiom)** Addition identity: Let *a* be a real number. $a + 0 = a$, $0 + a = a$
Conjecture Goldbach's conjecture: Every even integer, except 2, is equal to the sum of two prime numbers.	

Review your notes and Chapter 12 by using the Chapter Review on pages 747–750 of your textbook.